In Memoriam, Solomon Marcus

In Memoriam, Solomon Marcus

Editors

Cristian S. Calude
Gheorghe Paun

MDPI • Basel • Beijing • Wuhan • Barcelona • Belgrade • Manchester • Tokyo • Cluj • Tianjin

Editors
Cristian S. Calude
University of Auckland
New Zealand

Gheorghe Paun
Institute of Mathematics of the
Romanian Academy
Romania

Editorial Office
MDPI
St. Alban-Anlage 66
4052 Basel, Switzerland

This is a reprint of articles from the Special Issue published online in the open access journal *Axioms* (ISSN 2075-1680) (available at: http://www.mdpi.com).

For citation purposes, cite each article independently as indicated on the article page online and as indicated below:

LastName, A.A.; LastName, B.B.; LastName, C.C. Article Title. *Journal Name* **Year**, *Volume Number*, Page Range.

ISBN 978-3-0365-3475-6 (Hbk)
ISBN 978-3-0365-3476-3 (PDF)

Cover image courtesy of Cristian S. Calude

Contents

About the Editors

Cristian S. Calude is a member of Academia Europaea, a Chair Professor of Computer Science and Director of the Centre for Discrete Mathematics and Theoretical Computer Science at the University of Auckland, New Zealand. He has results in computational complexity, algorithmic information theory, quantum theory, history and philosophy of mathematics and computer science. Websites: http://en.wikipedia.org/wiki/Cristian_S._Calude, https://www.cs.auckland.ac.nz/%7Ecristian/awards.html

Gheorghe Paun is a member of the Romanian Academy and of Academia Europaea, and has now retired from the Institute of Mathematics of the Romanian Academy. Paun's research interests lie in formal language theory, DNA computing, and membrane computing.

Preface to "In Memoriam, Solomon Marcus"

Solomon Marcus (https://en.wikipedia.org/wiki/Solomon_Marcus) was a Romanian mathematician, member of the Romanian Academy, and professor at the University of Bucharest. He was a polymath with research in mathematics (mathematical analysis, measure theory, topology, mathematical and computational linguistics), theoretical computer science, poetics, linguistics, semiotics, philosophy, and history of science and education.

This book commemorates Marcus's fifth death anniversary with a selection of articles in mathematics, theoretical computer science, and physics written by authors who work in Marcus's research fields, some of whom have been influenced by his results and/or have collaborated with him. Marcus's (currently) final mathematical paper was published in *Axioms*, https://www.mdpi.com/2075-1680/7/1/15.

This book includes ten papers, one reviewing the contributions of Solomon Marcus to Theoretical Computer Science and Applications, and the others devoted to Membrane Computing, Interdimensionality, Network Topologies, Turing Machines, Complexity of Quasiperiodic Infinite Words, Kolmogorov Complexity, Cover Automata, and Communication Patterns.

Cristian S. Calude , Gheorghe Paun
Editors

axioms MDPI

Commentary

Solomon Marcus Contributions to Theoretical Computer Science and Applications

Cristian S. Calude [1,*,†] and Gheorghe Păun [2,†]

1 School of Computer Science, University of Auckland, 92019 Auckland, New Zealand
2 Institute of Mathematics, Romanian Academy, 014700 Bucharest, Romania; curteadelaarges@gmail.com
* Correspondence: cristian@cs.auckland.ac.nz
† These authors contributed equally to this work.

Abstract: Solomon Marcus (1925–2016) was one of the founders of the Romanian theoretical computer science. His pioneering contributions to automata and formal language theories, mathematical linguistics and natural computing have been widely recognised internationally. In this paper we briefly present his publications in theoretical computer science and related areas, which consist in almost ninety papers. Finally we present a selection of ten Marcus books in these areas.

Keywords: automata theory; formal language theory; bio-informatics; recursive function theory

Citation: Calude, C.S.; Păun, G. Solomon Marcus Contributions to Theoretical Computer Science and Applications. *Axioms* **2021**, *10*, 54. https://doi.org/10.3390/axioms10020054

Academic Editor: Florin Felix Nichita

Received: 4 March 2021
Accepted: 30 March 2021
Published: 5 April 2021

1. Introduction

In 2005, on the occasion of the 80th birthday anniversary of Professor Solomon Marcus, the editors of the present volume, both his disciples, together with his friend Professor G. Rozenberg, from Leiden, The Netherlands, have edited a special issue of *Fundamenta Informaticae* (vol. 64), with the title *Contagious Creativity*. This syntagma describes accurately the activity and the character of Marcus, a Renaissance-like personality, with remarkable contributions to several research areas (mathematical analysis, mathematical linguistics, theoretical computer science, semiotics, applications of all these in various areas, history and philosophy of science, education), with many disciples in Romania and abroad and with a wide recognition all around the world. Marcus projected his mathematical thinking in all domains in which he worked. Here is an example from semiotics, in the words of the Finnish musicologist and semiologist E. Tarasti (President of the International Association for Semiotic Studies (2004–2014), see N.-S. Drăgan, In Memoriam Solomon Marcus, "Hide and seek" with Solomon Marcus and Umberto Eco, *Book of Abstracts*, First Edition of the International Conference Semiosis in Communication, 1–3, Bucharest, 2016.):

> No other semiotician is so accurate and challenging in his reasoning about fundamental issues of our discipline.

In what follows we only briefly describe his contributions to theoretical computer science and related areas, especially to automata and formal language theories, natural computing (DNA and membrane computing), applications of grammars in various domains, recursive function theory and provability in mathematics, as well as a selection of his many books in these areas. Some re-printed in S. Marcus, *Words and Languages Everywhere*, Polimetrica, Milano, 2007, but almost all collected in the two-volume book G. Păun (ed.), *Solomon Marcus, Selected Papers—Computer Science*, Spandugino Publ. House, Bucharest, 2018, abbreviated SPCS. Our choices have been guided by *SPCS*.

Marcus' pioneering book *Gramatici şi automate finite (Grammars and Finite Automata)*, published in 1964 in Romanian is one of the first monographs in the world on this subject. This book, written in a rigorous mathematical language at a time when the domain was in infancy, covers automata and language theories, closely linking finite automata and Chomsky regular grammars. The book ends with a chapter on the relations between natural languages and regular grammars, a theme which motivated Marcus' interest and his many publications in mathematical linguistics. Unfortunately, the book, written in Romanian, was not translated into any other language; hence, it remained almost unknown internationally. This is not the case with many of his subsequent books, specifically those in mathematical linguistics, some of which will be listed in this paper. These books have been translated in several languages (French, English, German, Russian, Italian, Czech, Spanish, Greek and other languages) and then published by Academic Press, Dunod, Nauka and other well-known international publishers. Without exception, they had a very high international audience and impact.

His first paper in formal language theory was published in 1963 and it is illustrative for his permanent interest in building bridges between apparently disjoint research areas; in this case, finite automata, regular grammars, arithmetical progressions. Symmetrically, his last paper, published 50 years later, returns to bio-informatics, a domain which he somehow prognosticated (too early) in the beginning of the 70's.

2. A Working Classification

It is difficult to classify the theoretical computer science papers of Marcus because of their inter/multi-disciplinarity. In *SPCS*, the papers have been classified into four large categories: Formal language theory, applications of formal language theory, bio-informatics, and recursive function theory. We will use this classification here too.

In the first class there are papers dealing with finite state grammars and automata, contextual grammars, the history of formal language theory, combinatorics on words and on infinite sequences (periodicity and quasi-periodicity, unavoidable patterns, density of words of a given length), mathematical analysis notions adapted to formal language theory, and so on.

The last category deserves a closer study, which we only suggest here: To systematically extend notions/ideas from mathematical analysis to formal language theory in general and to combinatorics on words in particular (a symmetric study is worth carrying out for applications of formal languages to other mathematical areas, e.g., number theory by classifying various classes of numbers in Chomsky's hierarchy, characterising them with grammars, etc.). This was a direction of research programmatically explored by Marcus. The title of his 1999 paper is explicit and significant in this respect: *From real analysis to discrete mathematics and back*, followed by details: *Symmetry, convexity, almost periodicity, and strange attractors*. In the beginning of this paper he wrote:

> Despite its importance, the relation between continuous and discrete mathematics is a rather neglected topic. (...) Working in real analysis in the fifties and in the sixties and then in discrete mathematics (the mathematical theory of languages), I became interested to look for the discrete analog of some facts belonging to continuous mathematics.

Among the most fruitful ideas of this kind we mention several variants of the Darboux property for languages, the basic one being the following: If we have three families of languages, $\mathcal{L}_1 \subset \mathcal{L}_2 \subset \mathcal{L}_3$, conceivably belonging to a larger hierarchy of families of languages, possibly infinite, and two languages $L_1 \in \mathcal{L}_1$, $L_3 \in \mathcal{L}_3 \setminus \mathcal{L}_2$, can we find a language $L_2 \in \mathcal{L}_2$ such that $L_1 \subset L_2 \subset L_3$? Various definitions of symmetry, attractors, periodicity, convexity, etc., have been extended to strings. In all cases, Marcus used to define a series of subtle variants, of the type left-, right-, almost-, pseudo-, weak-, strong-, etc. Marcus had an unbounded creativity to pose open problems, and these papers never missed them; quite a few papers solved such problems, some of them with Marcus as a coauthor.

Actually, formulating open problems and suggesting research directions is one of the specific features of "Marcus' style". Many of the questions formulated by Marcus were addressed by his disciples, collaborators, by researchers in mathematics and computer science from Romania and other countries. Some problems were, partially or totally, solved—many of them are still waiting for solutions.

3. A Constant Interest for Bio-Informatics

We mentioned before that in the 1970s Marcus published "too early" a paper dealing with applications of mathematical linguistics and formal language theory in biology, specifically in the genomics area. The year was 1974 and the title of the paper is *Linguistic structures and generative devices in molecular genetics.*

Bio-informatics can be understood in two senses, as an attempt to use computer science in biology, providing notions, tools, techniques to the biologist and, mainly in the last decades, in the opposite direction, to utilise ideas inspired from biology in developing algorithms in computer science, and in hardware too, as is the case in DNA computing—DNA molecules do computations. In his paper, Marcus considered both directions. In the first direction of research he synthesised previous approaches and results; in the second one he proposed new research vistas for using mathematical (linguistic) tools in addressing questions in the genetic area, to model the DNA and its biochemistry. Speculations about using DNA molecules as a support for computations were published only later (by M. Conrad, R. Feynman, C. H. Bennet), while the first computing model based on an operation specific to DNA recombination was introduced only in 1987 by T. Head (another friend of Marcus). However, it is worth emphasising the attention paid by Marcus, in this first paper and also in many others, to a 1965 proposal formulated by the Polish mathematician Z. Pawlak (famous for introducing in early 1990s, the rough sets), to generate proteins starting from amino acids; the method used a specific representation of amino acids and certain picture grammars. (This is the reason Marcus considered Z. Pawlak a precursor of picture grammars, a type of generative mechanisms developed later.)

Over the years, Marcus was constantly interested in the (mathematical) linguistic approach to cellular biology, to applications in genomics and life sciences. For instance, after the apparition of DNA computing in 1994, and especially after the initiation of membrane computing in 1998, he had contributed to these areas with a series of papers and participated to several international meetings dedicated to these subjects, in Romania and abroad. As expected, the inter-disciplinary approach, typical to Marcus, is always present in his contributions—here are two illustrative titles of papers in membrane computing, *Membranes versus DNA* and *Bridging P systems and genomics*, presented at the first meetings devoted to membrane computing (Curtea de Argeş, Romania, 2001, 2002). Actually, in 2002, he proposed a slogan which became folklore in this research area:

Life = DNA software + membrane hardware.

As expected, in this area too he proposed several research directions, some of them truly "non-standard" ("too" inter-disciplinary) at the first sight. We only cite two examples of ideas not yet explored: To consider membranes with a topology different from the usual one (vesicle-like membranes), where the separation between inside and outside is crisp (for example, to study membranes similar to Klein's bottle), and, respectively, to use multisets, the sets with a multiplicity associated with their elements (the usual data structure in membrane computing) described by Pawlak rough sets.

4. Marcus Contextual Grammars

In a paper simply called "Contextual grammars" (published in 1969 in *Revue Roumaine de Mathématiques Pures et Appliquées*) Marcus has introduced the grammars which are now called Marcus contextual grammars, a branch of formal language theory. In fact, the paper was presented one year before in an international linguistics conference held in Stockholm, Sweden.

The paper has ten pages, but currently there probably exist more than 400 papers on contextual grammars, about two dozen of PhD and Master Theses, as well as two monographs, one published by the Publishing House of the Romanian Academy, Bucharest, 1982 (in Romanian), and one by Kluwer Publishing, The Netherlands, in 1997 (Marcus Contextual Grammars), both of them authored by Gh. Păun. In the second volume of the massive *Handbook of Formal Languages*, Springer-Verlag, 1997 (three volumes), edited by G. Rozenberg and A. Salomaa, there are two chapters dedicated to this topic, one by

Marcus, "Contextual grammars and natural languages", which discusses motivations and developments in this area, and another more technical one, "Contextual grammars and formal languages", by A. Ehrenfeucht, Gh. Păun, and G. Rozenberg.

The idea has the origins in algebraic linguistics: For a natural language L (over an alphabet V), with every word w over V one associates a set of contexts $\langle u, v \rangle$ over V which accept w with respect to L (that is, $uwv \in L$). Can we use this process of selecting words by contexts, in order to describe a language? One can also conversely state it. The answer was initially given in the form of simple contextual grammars, triples of the form $G = (V, A, C)$, where V is an alphabet, A is a finite language over V (its elements are called axioms), and C is a finite set of contexts over V. Such a grammar generates a language $L(G)$ which contains (1) all axioms in A and (2) all strings obtained from axioms by adjoining contexts to them. More formally, $L(G)$ contains all strings of the form $u_n \dots u_1 x v_1 \dots v_n$, where $x \in A$ and $\langle u_i, v_i \rangle \in C$ for all $1 \leq i \leq n$, with $n \geq 0$; for $n = 0$ the string is an axiom from A.

This simple model does not have a powerful generative capacity. Moreover, it does not take into consideration the string-contexts selectivity mentioned above. However, at the end of the paper, Marcus also proposes the contextual grammars with choice, $G = (V, A, C, \varphi)$, where $\varphi : V^* \to 2^C$ is the selection mapping (of contexts by the strings). This time, a string is in $L(G)$ if it is of the form $u_n \dots u_1 x v_1 \dots v_n$ as above with $x \in A$, $\langle u_1, v_1 \rangle \in \varphi(x)$, and $\langle u_i, v_i \rangle \in \varphi(u_{i-1} \dots u_1 x v_1 \dots u_{i-1})$ for all $i = 2, \dots, n$.

A great research program started from there, following the usual questions of formal language theory: Variants (extensions and restrictions), characterisations, generative power, comparisons of the obtained families among them and with the known families of languages, especially with those in the Chomsky hierarchy, closure and decidability properties, parsing complexity, equivalent automata, etc.

An important detail, which makes Marcus contextual grammars so attractive is the fact that they are not using, like the Chomsky grammars, nonterminal symbols, categorial auxiliary symbols: They are intrinsic grammars as each derived string belongs to the generated language.

Still, there was an embarrassing restriction in the initial model, the possibility to adjoin contexts only in the ends of the current string. A real breakthrough was proposed at the end of the 1970s, when the Vietnamese Nguyen Xuan My came to Romania to start a PhD with Marcus. In a joint paper Nguyen-Păun, the inner contextual grammars have been introduced: The contexts can be added in any place inside the current string, under the control of the selection mapping. (Formally, an inner contextual grammar is a usual contextual grammar with choice, $G = (V, A, C, \varphi)$, with $\varphi : V^* \to 2^C$, with the language $L(G)$ defined as the smallest language $L \subseteq V^*$ such that (i) $A \subseteq L$ and (ii) if $x_1 x_2 x_3 \in L$ and $\langle u, v \rangle \in \varphi(x_2)$, then $x_1 u x_2 v x_3 \in L$.) In this way, the generative capacity has significantly increased, the flexibility (hence the adequacy) of the model has been accordingly augmented.

Another important advance in this area was made at the beginning of the 1990s, when G. Rozenberg, A. Salomaa, A. Ehrenfeucht became interested in contextual grammars. Details can be found in Kluwer's monograph mentioned before and in two chapters in the *Handbook of Formal Languages*.

Progress was rather rapid. Certain classes of contextual grammars have been proved to be relevant for modelling typical constructions in natural languages (duplication, multiple agreements, crossed agreements) and classes of contextual grammars which are mildly context sensitive in the sense requested by linguists (A. K. Joshi and others) have been introduced. They are parsable in polynomial time and contain strings whose lengths do not make large jumps—sometimes one asks only that the language be semilinear.

In this way, the impressive bibliography we mentioned above has been accumulated—and this bibliography is still growing.

5. Applications of Formal Language Theory

In this class we have included the papers devoted to applications of grammars and automata. This was a really central and continuous interest of Marcus, also passed onto his students and collaborators. The domains of applicability are very diverse: Natural and programming languages, the semiotics of folklore fairy tales, the modelling of economic processes, diplomatic negotiations, the medical diagnosis, the semiotics of theatre, action theory, learning theory, chemistry, genetics.

These applications should be placed in a more general context under the slogan linguistics as a pilot-science, a catchphrase coined by C. Levi-Strauss: Adopted, extended and transformed by Marcus it became a real research program for his Romanian school of mathematical linguistics and formal language theory.

The grounding assumption, also explored by M. Nowakowska in her book *Languages of Action, Languages of Motivations*, Mouton, The Hague, 1973, was that many processes/activities can be described as sequences of elementary actions ("semantic marks"), sequences which are governed by precise restrictions which can be described by syntactic rules. Thus, languages describing actions and grammars describing languages of actions came into stage. Combined with the Chomskian hypothesis that the linguistic competence is innate and influences all other competences of the human brain, Levi-Strauss's slogan became Marcus' formal linguistics as a pilot-science. Indeed, a large variety of processes, from fairy tales description to economic processes proved to be described, at convenient levels of abstraction, by grammars of the types initially developed in linguistics.

6. Recursive Function Theory and Provability

The last category of papers we mention deals with recursive functions and provability in mathematics; it contains fewer papers, but some of these papers have a special significance, as they clarify an important paternity in the history of computability. Specifically, they proved that the first example of a recursive function which is not primitive recursive was constructed by G. Sudan in 1927, simultaneously with and independently of W. Ackermann, who was credited before with this achievement (1928). The problem was examined by Marcus in collaboration with C. Calude and I. Ţevy, following a suggestion coming from G. C. Moisil.

It is important to mention that Marcus was constantly concerned with adequately valuing the history of the Romanian mathematics: Pointing out the priorities in this area was already one of the main goals of his well-known book *Din gândirea matematică românească (From the Romanian Mathematical Thinking)*, Scientific and Encyclopaedic Publishing House, Bucharest, 1975.

This group also includes a few papers on provability in mathematics, at different levels of formalisation and with various tools, including proof-assistants.

7. Papers

7.1. Formal Language Theory

1. S. Marcus, Automates finis, progressions arithmétiques et grammaires à un nombre fini d'etats. *Comptes rendus de l'Academie des Sciences Paris*, 256, 17 (1963), 3571–3574.
2. S. Marcus, Sur un modéle de H. B. Curry pour le langage mathématique. *Comptes rendus de l'Academie des Sciences Paris*, 258, 7 (1964), 1954–1956.
3. S. Marcus, Sur les grammaires à un nombre fini d'états. *Cahiers de Linguistique Théorique et Appliquée*, 2 (1965), 146–164.
4. S. Marcus, Analytique et génératif dans la linguistique algébrique. In *To Honor Roman Jakobson II*, Mouton, The Hague, 1967, 1252–1261.
5. S. Marcus, Contextual grammars. *Revue Roumaine de Mathématiques Pures et Appliquée*, 14, 10 (1969), 1525–1534; also, Preprint nr. 48, Intern. Conf. Comput. Ling., Stockholm, 1968.
6. S. Marcus, Deux types nouveaux de grammaires génératives. *Cahiers de Linguistique Théorique et Appliquées*, 6 (1969), 67–74.

7. S. Marcus, Darboux property and formal languages. *Revue Roumaine de Mathématiques Pures et Appliquées*, 22, 10 (1977), 1449–1451.
8. S. Marcus, Problems. *Bulletin of the European Association for Theoretical Computer Science*, 27 (1985), 245.
9. S. Marcus, Formal languages before Axel Thue? *Bulletin of the European Association for Theoretical Computer Science*, 34 (1988), 62.
10. S. Marcus, Din istoria limbajelor formale. *Al doilea Colocviu Naţional de Limbaje, Logică, Lingvistică Matematică*, Braşov, iunie 1888, 1–9.
11. S. Marcus, Gh. Păun, Langford strings, formal languages and contextual ambiguity, *Intern. J. Computer Math.*, 26, 3 + 4 (1989), 179–191.
12. L. Kari, S. Marcus, Gh. Păun, A. Salomaa, In the prehistory of formal languages, Gauss languages. *Bulletin EATCS*, 46 (1992), 124–139.
13. S. Marcus, Fivefold symmetry: A generative approach. In *Caiet de Semiotică*. Univ. Timişoara, 9 (1992), 1–23.
14. S. Marcus, Thirty-six years ago. The beginning of the formal language theory. In *Salodays in Theoretical Computer Science*, May 1992 (A. Atanasiu, C.S. Calude, eds.), Univ. Hyperion, Bucharest, 1993.
15. S. Marcus, Symbols in a multidimensional space. In *SEMIOTICS 1990* (K. Haworth, J. Deely, T. Prewitt, eds.) with *SYMBOLICITY* (J. Bernard, J. Deely, V. Voigt, G. Withalm, eds.), The Semiotic Soc. of America, 1993, 115–126.
16. J. Dassow, S. Marcus, Gh. Păun, Iterated reading of numbers and "black-holes". *Periodica Mathematica Hungarica*, 27, 2 (1993), 137–152.
17. J. Dassow, S. Marcus, Gh. Păun, Iterative reading of numbers; Parikh mappings, parallel rewriting, infinite sequences. *Preprint of. Tech. Univ. Otto von Guericke Univ., Magdeburg*, July 1993, 18 pp.
18. J. Dassow, S. Marcus, Gh. Păun, Iterative reading of numbers: The ordered case. In *Developments in Language Theory. At the Crossroad of Mathematics, Computer Science and Biology* (G. Rozenberg, A. Salomaa, eds.), World Sci. Publ., Singapore, 1994, 157–168.
19. S. Marcus, Gh. Păun, On symmetry in languages. *Intern. J. Computer Math.*, 52, 1/2 (1994), 1–15.
20. S. Marcus, Gh. Păun, Infinite words and their associated formal languages. In *Salodays in Auckland* (C. Calude, M.J.J. Lennon, H. Maurer, eds.), Auckland Univ. Press, 1994, 95–99.
21. S. Marcus, Al. Mateescu, Gh. Păun, A. Salomaa, On symmetry in strings, sequences and languages. *Intern. J. Computer Math.*, 54, 1/2 (1994), 1–13.
22. S. Marcus, Gh. Păun, Infinite (almost periodic) words, formal languages, and dynamical systems. *Bulletin EATCS*, 54 (1994), 224–231.
23. M. Kudlek, S. Marcus, A. Mateescu, Contextual grammars with distributed catenation and shuffle. *Found. of Computation Theory, FCT*, LNCS 1279 (B.S. Chlebus, L. Czeja, eds.), Springer, Berlin, 1997, 269–280.
24. J. Dassow, S. Marcus, Gh. Păun, Convex and anti-convex languages. *Intern. J. Computer Math.*, 69, 1-2 (1998), 1–16.
25. S. Marcus, C. Martin-Vide, Gh. Păun, On the power of internal contextual grammars with maximal use of selectors. *Conf. Automata and Formal Languages*, Salgotarjan, 1996, *Publicationes Mathematicae, Debrecen*, 54 (1999), 933–947.
26. S. Marcus, On the length of words. In *Jewels are Forever. Contributions on Theoretical Computer Science in Honor of Arto Salomaa* (J. Karhumaki, H. Maurer, Gh. Păun, G. Rozenberg, eds.), Springer, Berlin, 1999, 194–203.
27. S. Marcus, From real analysis to discrete mathematics and back: symmetry, convexity, almost periodicity and strange attractors. *Real Analysis Exchange*, 25, 1 (1999-2000), 125–128.
28. S. Marcus, Pseudo-slender languages and their infinite hierarchy. *Ninth Intern. Conf. Automata and Formal Languages*, Vasscecseny, Hungary, August 1999, 1-2.

29. S. Marcus, C. Martin-Vide, V. Mitrana, Gh. Păun, A new–old class of linguistically motivated regulated grammars. *Computational Linguistics in the Netherlands, 2000*, Rodopi, New York, 2001, 111–125.
30. S. Marcus, Bridging two hierarchies of infinite words. *Journal of Universal Computer Sci.*, 8, 2 (2002), 292–296.
31. S. Marcus, Quasiperiodic infinite words. *Bulletin EATCS*, 82 (2004), 170–174.
32. L. Ilie, I. Petre, S. Marcus, Periodic and Sturmian languages. *Information Processing Letters*, 98, 6 (2006), 242–246.
33. S. Marcus, Mild context-sensitivity, after twenty years. *Fundamenta Informaticae*, 73, 1/2 (2006), 203–204.
34. T. Monteil, S. Marcus, Quasiperiodic words: multi-scale case and dynamical properties. `https://arxiv.org/abs/math/0603354`, March 2006.
35. P. Dömösi, M. Ito, S. Marcus, Marcus contextual languages consisting of primitive words. *Discrete Mathematics*, 308 (2008), 4877–4881.

7.2. Applications of Formal Languages

1. S. Marcus, Linguistique générative, modèles analytiques et linguistique générale. *Revue Roumaine de Linguistique*, 14, 4 (1969), 313–326.
2. S. Marcus, Linguistics for programming languages. *Revue Roumaine de Linguistique. Cahiers de Linguistique Théorique et Appliquée*, 16, 1 (1970), 29–39.
3. S. Fotino, S. Marcus, Gramatica basmului (I). *Revista de Etnografie şi Folclor*, 18, 4 (1973), 255–277.
4. S. Fotino, S. Marcus, Gramatica basmului (II). *Revista de Etnografie şi Folclor*, 18, 5 (1973), 349–363.
5. E. Celan, S. Marcus, Le diagnostique comme langage (I). *Cahiers de Linguistique Théorique et Appliquée*, 10, 2 (1973), 163–173.
6. S. Marcus, Linguistics as a pilot science. In *Current Trends in Linguistics* (Th. Sebeok, ed.), Mouton, The Hague, 1974, 2871–2887, şi în *Studii şi cercetări lingvistice*, 20, 3 (1969), 235–245.
7. S. Marcus, Applications de la théorie des langages formels en économie et organisation, *Cahiers de Linguistique Théorique et Appliquée*, 13, 2 (1976), 583–594. Also published in *Annales de la Faculté des sciences de l'Université Nationale de Zaïre*, Kinshasa, vol. 3, 1977, nr. 1, p. 125–147
8. S. Marcus, Languages, grammars and negotiations. Some suggestions. In *Mathematical Approaches to International Relations* (M. Bunge, J. Galtung, M. Maliţa, eds.), vol. 2, Romanian Acad. of Social and Political Sci., Bucharest, 1977, 378–385.
9. S. Marcus, A new generative approach to fairy-tales. *Ethnologica*, annexe á la publication *Recherches sur l'histoire comparative des constitutions et du droit*, Bucharest, 1978, 14–17.
10. C. Calude, S. Marcus, Gh. Păun, The universal grammar as a hypothetical brain, *Rev. Roumaine Ling.* 25, 5 (1979), 479–489.
11. S. Marcus, Lingvistica şi logica. *Studii şi cercetări lingvistice*, 30, 3 (1979), 247–249.
12. Al. Balaban, M. Barasch, S. Marcus, Computer programs for the recognition of acyclic regular isoprenoid structures. *MATCH - Mathematical Chemistry*, 5 (1979), 239–261.
13. S. Marcus, Learning, as a generative process. *Revue Roumaine de Linguistique*, 24 (1979), *Cahiers de Linguistique Théorique et Appliquée*, 16, 2 (1979), 117–130.
14. S. Marcus, Semiotics of theatre: A mathematical linguistic approach, *Revue roumaine de linguistique*, 25, 3 (1980), 161–189.
15. Al. Balaban, M. Barasch, S. Marcus, Picture grammars in Chemistry. Generation of acyclic isoprenoid structures. *MATCH - Mathematical Chemistry*, 8 (1980), 193–213.
16. Al. Balaban, M. Barasch, S. Marcus, Computer program for the recognition of standard isoprenoid structures. *MATCH - Mathematical Chemistry*, 8 (1980), 215–268.

17. Al. Balaban, M. Barasch, S. Marcus, Codification of acyclic isoprenoid structures using context-free grammars and pushdown automata. *MATCH - Mathematical Chemistry*, 12 (1981), 25–64.
18. S. Marcus, La lecture générative. *Degrés*, 28 (1981), 61–66.
19. S. Marcus, Limbaje naturale şi limbaje artificiale. *Lucrările primului Colocviu Naţional de Limbaje, Logică, Lingvistică Matematică*, Braşov, iunie 1886, 1–18.
20. S. Marcus, Interplay of innate and acquired in some mathematical models of language learning. *Revue Roumaine de Linguistique*, 34, 2 (1989), 101–116.
21. S. Marcus, Semiotics and formal artificial languages. In *Encyclopaedia of Computer Science and Technology* (A. Kent, J.G. Williams, eds.), vol. 29, Marcel Dekker Inc., New York, 1994, 393–405.
22. S. Marcus, C. Martin-Vide, Gh. Păun, Contextual grammars versus natural languages. *Speech and Computers Conf., SPECOM 96*, St. Petersburg, 1996, 28–33.
23. S. Marcus, Contextual grammars and natural languages. In *Handbook of Formal Languages* (G. Rozenberg, A. Salomaa, eds.), vol. II, Springer, Berlin, 1997, 215–235.
24. S. Marcus, C. Martin-Vide, Gh. Păun, Contextual grammars as generative models of natural languages. *Fourth Meeting on Mathematics of Language, MOL 4*, Philadelphia, 1995, *Computational Linguistics*, 24, 2 (1998), 245–274.
25. S. Marcus, Linguistic and semiotic preliminaries of contextual languages. *Math. and Comput. Analysis of Natural Languages. Proc. Second Intern. Conf. on Math. Linguistics* (C. Martin-Vide, ed.), Tarragona, 1996, John Benjamins, Amsterdam, 1998, 47–57.
26. S. Marcus, Contextual grammars, learning processes and the Kolmogorov-Chaitin metaphor. *Math. Found. Computer Sci. Workshop on Mathematical Linguistics*, Brno, August 1998, Bericht 213, Univ. Hamburg, Juli 1998, 1–12.
27. S. Marcus, Reading numbers as a metaphor of the universe. In *BRIDGES - Math. Connections in Art, Music and Science*, Southwestern College, Winfield, Kansas, 1999 (R. Sarhangi, ed.), Gilliland Printing, Maryland, 1999, 302.
28. S. Marcus, History as text: Xenopol's series between structuralism and generative formal grammars. *Romanian J. Information Sci. and Technology*, 5, 1-2 (2002), 5–8.
29. S. Marcus, Formal languages: Foundations, prehistory, sources, and applications. In *Formal Languages and Applications* (C. Martín-Vide, V. Mitrana, Gh. Păun, eds.), Springer, Berlin, 2004, 11–53.

7.3. Bio-informatics

1. S. Marcus, Linguistic structures and generative devices in molecular genetics. *Cahiers de Linguistique Théorique et Appliquée*, 11, 1 (1974), 77–104.
2. C. Calude, S. Marcus, Gh. Păun, The universal grammar as a hypothetical brain. *Revue Roumaine de Linguistique*, 24, 5 (1979), 479–489.
3. S. Marcus, Hidden grammars. In *Developments in Language Theory* (G. Rozenberg, A. Salomaa, eds.), World Sci. Publ., Singapore, 1994, 453–460.
4. S. Marcus, Language, at the crossroad of computation and biology. In *Computing with Biomolecules. Theory and Experiments* (Gh. Păun, ed.), Springer, Singapore, 1998, 1–34.
5. S. Marcus, Bags and beyond them. *Pre-proc. Workshop on Multiset Processing*, Curtea de Argeş, Romania, 21-25 August 2000, Report CDMTCS-140, C.S. Calude, M.J. Dinneen, Gh. Păun, eds., 191–192.
6. S. Marcus, Tolerance multisets. In *Multiset Processing. Mathematical, Computer Science and Molecular Computing Points of View* (C.S. Calude, Gh. Păun, G. Rozenberg, A. Salomaa, eds.), LNCS 2235, Springer, Berlin, 2001, 217–223.
7. S. Marcus, Membranes versus DNA. *Pre-proc. Workshop on Membrane Computing* (WMC-CdA 2001) (C. Martin-Vide, Gh. Păun, eds.), Rovira i Virgili Univ., Tarragona, Spain, 2001, 193–198, and *Fundamenta Informaticae*, 49, 1-3 (2002), 223–227.
8. S. Marcus, Bridging P systems and genomics. In *Membrane Computing. International Workshop, WMC-CdeA 2002, Curtea de Argeş, Romania, August 2002, Revised Papers* (Gh.

Păun, G. Rozenberg, A. Salomaa, C. Zandron, eds.), LNCS 2597, Springer, Berlin, 2003, 371–376.
9. S. Marcus, Symmetry phenomena in infinite words, with biological, philosophical and aesthetic relevance. *Symmetry: Culture and Science*, 14/15 (2003-2004), 477–487.
10. S. Marcus, The duality of patterning in molecular genetics. In *Aspects of Molecular Computing. Essays Dedicated to Tom Head on the Occasion of His 70th Birthday* (N. Jonoska, Gh. Păun, eds.), LNCS 2950, Springer, Berlin, 2004, 318–321.
11. S. Marcus, Z. Pawlak, a precursor of DNA computing and of picture grammars. *Fundamenta Informaticae*, 75, 1/4 (2007), 331–334.
12. G. Ciobanu, S. Marcus, Gh. Păun, New strategies of using the rules of a P system in a maximal way. Power and complexity. *Romanian J. Information Sci. and Technology*, 12, 2 (2009), 157–173.
13. S. Marcus, The biological cell in spectacle. In *Membrane Computing. 10th Intern. Workshop, WMC 2009, Curtea de Argeş, Romania, August 2009, Revised Selected and Invited Papers* (Gh. Păun, M.J. Pérez-Jiménez, A. Riscos-Núñez, G. Rozenberg, A. Salomaa, eds.), LNCS 5957, Springer, Berlin, 2010, 95–103.
14. S. Istrail, S. Marcus, Alan Turing and John von Neumann—Their Brains and Their Computers. *Membrane Computing, 13th Intern. Conf., Budapest, August 2012, Revised Selected Papers* (E. Csuhaj-Varjú, M. Gheorghe, G. Rozenberg, A. Salomaa, G. Vaszil, eds.), LNCS 7762, Springer, Berlin, 2013, 26–35.

7.4. Recursive Function Theory and Provability

1. C. Calude, S. Marcus, I. Ţevy, Sur les fonctions récursives qui ne sont pas récursives primitives, *Revue Roumaine des Sciences Sociales*, Série de Philosophie et Logique 19, 3 (1975), 185–188.
2. C. Calude, S. Marcus, I. Ţevy, The first example of a recursive function which is not primitive recursive, *Historia Mathematica* 6 (1979), 380–384.
3. C. Calude, S. Marcus, I. Ţevy, Recursive properties of Sudan function, *Revue Roum. Math. Pures Appl.*, 25, 4 (1980), 503–507.
4. C. Calude, S. Marcus, Sudan's recursive but not primitive recursive function: A retrospective look, *Analele Universităţii din Bucureşti, Matematică-Informatică*, 38, 2 (1989), 25–30.
5. C. S. Calude, S. Marcus, L. Staiger, A topological characterization of random sequences, *Inform. Process. Lett.* 88 (2003), 245–250.
6. C. S. Calude, E. Calude, S. Marcus, Passages of proof, *Bull. Eur. Assoc. Theor. Comput. Sci.* 84 (2004), 167–188.
7. C. S. Calude, E. Calude, S. Marcus, Proving and programming, in C. S. Calude (ed.), *Randomness & Complexity, from Leibniz to Chaitin*, World Scientific, Singapore, 2007, 310–321.
8. C. S. Calude, S. Marcus, Mathematical proofs at a crossroad? in J. Karhumäki, H. Maurer, Gh. Păun, G. Rozenberg (eds.), *Theory Is Forever*, Lectures Notes in Comput. Sci. 3113, Springer-Verlag, Berlin, 2004, 15–28.
9. S. Marcus, S. M. Watt, What is an equation? *Proc. 14th International Symposium on Symbolic and Numeric Algorithms for Scientific Computing*, IEEE (2012), 23–29.
10. S. Marcus, Proofs and mistakes: Their syntactics, semantics, and pragmatics, *Semiotica* 188 (2012), 139–155.

8. Selected Books

1. S. Marcus, *Lingvistica matematică. Modele matematice în lingvistică* (*Mathematical Linguistics. Mathematical Models in Linguistics*), Ed. Didactică şi Pedagogică, Bucureşti, 1963. (In Romanian)
2. S. Marcus, *Gramatici şi automate finite (Grammars and Finite Automata)*, Ed. Academiei, Bucureşti, 1964. (In Romanian)

3. S. Marcus, *Introduction mathematique a la linguistique structurale*, Dunod, Paris, 1967. (In French)
4. S. Marcus, *Algebraic Linguistics; Analytical Models*, Academic Press, New York, 1967.
5. S. Marcus, *Introduzione alla linguistica matematica*, Casa editrice Riccardo Patron, Bologna, 1970. (with E. Nicolau and S. Stati) (in Italian)
6. S. Marcus, *Teoretiko-mnozestvennye modeli jazykov*, Ed. Nauka, Moscova, 1970. (In Russian)
7. S. Marcus (coordinator), *La sémiotique formelle du folklore. Approche linguistico-mathématiq* Ed. Klincksieck, Paris—Ed. Academiei, Bucureşti, 1978. (In French)
8. S. Marcus, *Mathematische Poetik*, Athenaeum Verlag, Frankfurt/Main, 1973. (In German)
9. S. Marcus, *Snmeia gia ta snmeia*, Ed. Pneumatikos, Atena, 1981. (In Greek)
10. S. Marcus, *Contextual Ambiguities in Natural and in Artificial Languages, Communication and Cognition*, Ghent, Belgium, vol.1, 1981; vol.2, 1983. (In German)

Author Contributions: Authors have contributed in equal parts. Both authors have read and agreed to the published version of the manuscript.

Funding: This research received no external funding.

Conflicts of Interest: The authors declare no conflict of interest.

axioms

MDPI

Article

Simulations between Network Topologies in Networks of Evolutionary Processors

José Ángel Sánchez Martín [1] and Victor Mitrana [1,2,*]

[1] Departamento de Sistemas Informáticos, Universidad Politécnica de Madrid, C/Alan Turing s/n, 28031 Madrid, Spain; joseangel.sanchez.martin@alumnos.upm.es
[2] National Institute for Research and Development of Biological Sciences, Independentei Bd. 296, 060031 Bucharest, Romania
* Correspondence: victor.mitrana@upm.es

Abstract: In this paper, we propose direct simulations between a given network of evolutionary processors with an arbitrary topology of the underlying graph and a network of evolutionary processors with underlying graphs—that is, a complete graph, a star graph and a grid graph, respectively. All of these simulations are time complexity preserving—namely, each computational step in the given network is simulated by a constant number of computational steps in the constructed network. These results might be used to efficiently convert a solution of a problem based on networks of evolutionary processors provided that the underlying graph of the solution is not desired.

Keywords: evolutionary processor; network of evolutionary processors; network topology; theory of computation; computational models

Citation: Sánchez Martín, J.Á.; Mitrana V. Simulations between Network Topologies in Networks of Evolutionary Processors. *Axioms* **2021**, *10*, 183. https://doi.org/10.3390/axioms10030183

Academic Editor: Cristian S. Calude

Received: 19 July 2021
Accepted: 6 August 2021
Published: 11 August 2021

1. Introduction

Networks of evolutionary processors (NEPs for short) have been extensively investigated in the last two decades since their generative variant has been introduced in [1]. An informal description of a NEP is as follows: it is a graph whose nodes are hosts for some very simple processors inspired by the basic mutations at the DNA nucleotide level, namely insertion, deletion, and substitution. Each processor is able to make just one of these operations on the data existing in the node that hosts it. Data may be organized as strings, multisets, two-dimensional pictures, graphs, etc. In this work, we consider that the data consist of strings. A very important assumption is that each string appears in an arbitrarily large number of identical copies such that if the processor can apply an operation to different sites of a string, the operation is actually applied simultaneously to each of these sites in different copies of the string. Furthermore, if more that one rule can be applied to a string, each rule is applied to a different copy of that string. This process described above is considered to be an evolutionary step. Each evolutionary step alternates with a communication step. In a communication step, all the strings that can leave a node (they can pass the output filter associated with that node) actually leave the node and copies of them enter each node connected to the left node, provided that they can pass the input filter of the arriving node. We say that an input string, which initially is in a designated node, called the input node, is accepted if another designated node, called the output node, is non-empty after a finite number of computational steps (evolution, communication). The complexity of a computation is defined in the usual way.

From the very beginning, NEPs have been proven to be computationally complete models [2,3], such that they have been used to solve hard problems [4]. Several variants have been considered depending on the positions of filters: filters associated with nodes (different filters [3], uniform filters [5], polarization [6]) or filters associated with edges [7]. Later on, several ways of simulating and implementing different variants of these networks have been reported [8–11]. A rather new and attractive direction of research has been to investigate the possibility of simulating directly and efficiently one variant by another without the intermediate step of an extra computational model (Turing machine, tag-system, register machine, etc.) in between, see, e.g., [5].

This work continues this line of research by proposing direct simulations between two NEPs such that the input one is an arbitrary NEP while the output one has a predefined topology that can be a complete graph, a star graph, or a grid. Thus, after a preliminary section with the basic definitions and concepts, we give the construction of a complete NEP equivalent to a given NEP. We continue with another section, where we give such a construction for a star graph and finally a construction for a grid NEP. A short conclusion ends the paper.

2. Basic Definitions

The basic concepts and notations that are to be used throughout the paper are defined in the sequel; the reader may consult [12] for basic concepts that are not defined here. We use the following concepts and notations:

- V^* is the set of all strings formed by symbols in V;
- $|x|$ is the length of string x;
- $\varepsilon \in V^*$ is the empty string, $|\varepsilon| = 0$;
- $alph(x)$ is the minimal alphabet V such that $x \in V^*$.

We now recall some definitions from a few papers where the networks of evolutionary processors have been introduced, see, e.g., [1], for the generating model, and [3,13,14], for the accepting model. Let $a \to b$ be a rule, where $a, b \in (V \cup \{\varepsilon\})$:

- If $a, b \in V$, then the rule is called a *substitution* rule;
- If $a \in V$ and $b = \varepsilon$, then the rule is called a *deletion* rule;
- If $a = \varepsilon$ and $b \in V$, then the rule is called an *insertion* rule.

The set of all substitution, deletion, and insertion rules over V is denoted by Sub_V, Del_V, and Ins_V, respectively.

Given a rule σ as above and a string $w \in V^*$, we define the following *actions* of σ on w , to any position ($*$), to the leftmost position (l), and to the rightmost position (r), as explained in the sequel:

– If $\sigma \equiv a \to b \in Sub_V$, then

$$\sigma^*(w) = \begin{cases} \{ubv : \exists u, v \in V^* \ (w = uav)\}, \\ \{w\}, \text{ otherwise} \end{cases}$$

According to this definition, applying a rule to a string may result in a finite number of strings. This implies that in our setting each string may appear in an arbitrarily large number of copies.

– If $\sigma \equiv a \to \varepsilon \in Del_V$, then $\sigma^*(w) = \begin{cases} \{uv : \exists u, v \in V^* \ (w = uav)\}, \\ \{w\}, \text{ otherwise} \end{cases}$

$$\sigma^r(w) = \begin{cases} \{u : w = ua\}, \\ \{w\}, \text{ otherwise} \end{cases} \qquad \sigma^l(w) = \begin{cases} \{v : w = av\}, \\ \{w\}, \text{ otherwise} \end{cases}$$

– If $\sigma \equiv \varepsilon \to a \in Ins_V$, then $\sigma^*(w) = \{uav : \exists u, v \in V^* \ (w = uv)\}$, $\sigma^r(w) = \{wa\}$, $\sigma^l(w) = \{aw\}$.

For every rule σ, $\alpha \in \{*, l, r\}$, and $L \subseteq V^*$, we define $\sigma^\alpha(L) = \bigcup_{w \in L} \sigma^\alpha(w)$. Given a finite and non-empty set of rules M, a string w and a language L, we define the followings:

$$M^\alpha(w) = \bigcup_{\sigma \in M} \sigma^\alpha(w) \text{ and } M^\alpha(L) = \bigcup_{w \in L} M^\alpha(w).$$

In the original papers mentioned above, the rewriting operations defined above were referred as *evolutionary operations* since they may be viewed as formal operations abstracted from local DNA mutations.

For two disjoint subsets P (permitting symbols) and F (forbidding symbols) of an alphabet V and a string z over V, we define the predicates:

$$\begin{array}{lll} \varphi^{(s)}(z;P,F) \equiv P \subseteq alph(z) & \wedge & F \cap alph(z) = \emptyset \\ \varphi^{(w)}(z;P,F) \equiv (P \neq \emptyset) \rightarrow (alph(z) \cap P \neq \emptyset) & \wedge & F \cap alph(z) = \emptyset. \end{array}$$

For every language $L \subseteq V^*$ and $\beta \in \{(s),(w)\}$, we define:

$$\varphi^{\beta}(L,P,F) = \{z \in L \mid \varphi^{\beta}(z;P,F)\}.$$

An *evolutionary processor* (EP) over an alphabet V is a tuple (M, PI, FI, PO, FO), where:

- M is a set of either substitution, or deletion or insertion rules over the alphabet V. Formally: $(M \subseteq Sub_V)$ or $(M \subseteq Del_V)$ or $(M \subseteq Ins_V)$. The set M represents the set of evolutionary rules of the processor;
- $PI, FI \subseteq V$ are the *input* permitting/forbidding symbols of the processor, while $PO, FO \subseteq V$ are the *output* permitting/forbidding symbols of the processor.

We denote the set of evolutionary processors over V by EP_V. A *network of evolutionary processors* (NEP for short) is a seven-tuple $\Gamma = (V, U, G, \mathcal{N}, \alpha, \beta, \underline{In}, \underline{Out})$, where:

- V and U are the input and network alphabets, respectively, $V \subseteq U$.
- $G = (X_G, E_G)$ is an undirected graph without loops, with the set of nodes X_G and the set of edges E_G. Each edge is given in the form of a binary set. G is called the *underlying graph* of the network;
- $\mathcal{N} : X_G \longrightarrow EP_U$ is a mapping which associates with each node $x \in X_G$ the evolutionary processor $\mathcal{N}(x) = (M_x, PI_x, FI_x, PO_x, FO_x)$;
- $\alpha : X_G \longrightarrow \{*, l, r\}$; $\alpha(x)$ gives the action mode of the rules of node x on the strings existing in that node;
- $\beta : X_G \longrightarrow \{(s),(w)\}$ defines the type of the *input/output filters* of a node. More precisely, for every node, $x \in X_G$, the following filters are defined:

  $$\begin{array}{r} \text{input filter: } \rho_x(\cdot) = \varphi^{\beta(x)}(\cdot; PI_x, FI_x), \\ \text{output filter: } \tau_x(\cdot) = \varphi^{\beta(x)}(\cdot; PO_x, FO_x). \end{array}$$

  That is, $\rho_x(z)$ (resp. $\tau_x(z)$) indicates whether or not the string z can pass the input (resp. output) filter of x. More generally, $\rho_x(L)$ (resp. $\tau_x(L)$) is the set of strings of L that can pass the input (resp. output) filter of x.
- $\underline{In}$ and $\underline{Out} \in X_G$ are the *input node*, and the *output node*, respectively, of the NEP.

A *configuration* of a NEP Γ as above is a function $C : X_G \longrightarrow 2^{U^*}$ which associates a multiset of strings $C(x)$ with every node x of Γ. As each string appears in an arbitrarily large number of copies, we work with the support of this multiset. For a string $w \in V^*$, we define the initial configuration of Γ on w by $C_0^{(w)}(\underline{In}) = \{w\}$ and $C_0^{(w)}(x) = \emptyset$ for all $x \in X_G \setminus \{\underline{In}\}$.

A configuration is followed by another configuration either by an *evolutionary step* or by a *communication step*. A configuration C' follows a configuration C by an evolutionary step if each component $C'(x)$, for some node x, is the result of applying all the evolutionary rules in the set M_x that can be applied to the strings in the set $C(x)$. Formally, configuration C' follows the configuration C by a an evolutionary step, written as $C \Longrightarrow C'$, if

$$C'(x) = M_x^{\alpha_x}(C(x)) \text{ for all } x \in X_G.$$

In a communication step of a NEP the following actions take place simultaneously for every node x:

(i) All the strings that can pass the output filter of a node are sent out of that node;

(ii) All the strings that left their nodes enter all the nodes connected to their original ones, provided that they can pass the input filter of the receiving nodes.

Note that, according to this definition, those strings that are sent out of a node and cannot pass the input filter of any node are lost.

Formally, a configuration C' follows a configuration C by a communication step (we write $C' \models C$) iff for all $x \in X_G$

$$C'(x) = (C(x) \setminus \tau_x(C(x))) \cup \bigcup_{\{x,y\} \in E_G} (\tau_y(C(y)) \cap \rho_x(C(y))).$$

Let Γ be a NEP, the computation of Γ on the input string $w \in V^*$ is a sequence of configurations $C_0^{(w)}, C_1^{(w)}, C_2^{(w)}, \ldots,$ where $C_0^{(w)}$ is the initial configuration of Γ on w, $C_{2i}^{(w)} \Longrightarrow C_{2i+1}^{(w)}$ and $C_{2i+1}^{(w)} \models C_{2i+2}^{(w)}$, by a for all $i \geq 0$. Note that the configurations are changed by alternative steps.

A computation as above *halts*, if there exists a configuration in which the set of strings existing in the output node <u>*Out*</u> is non-empty. Given a NEP Γ and an input string w, we say that Γ accepts w if the computation of Γ on w halts. Consequently, we define the *language accepted* by Γ by

$$L(\Gamma) = \{z \in V^* \mid \text{ the computation of } \Gamma \text{ on } z \text{ halts}\}.$$

The *time complexity* of the halting computation $C_0^{(z)}, C_1^{(z)}, C_2^{(z)}, \ldots C_m^{(z)}$ of Γ on $z \in V^*$ is denoted by $time_\Gamma(z)$ and equals m. The time complexity of Γ is the function from $\mathbb{N}$ to $\mathbb{N}$, $Time_\Gamma(n) = \max\{time_\Gamma(z) \mid z \in L(\Gamma), |z| = n\}$. In other words, $Time_\Gamma(n)$ delivers the maximal number of computational steps carried out by Γ for accepting an input string of length n.

3. Simulating Any NEP with a Complete NEP

Theorem 1. *Given an arbitrary NEP Γ, there exists a complete NEP Γ' such that the following two conditions are satisfied:*

1. $L(\Gamma) = L(\Gamma')$;
2. $Time_{\Gamma'}(n) \in \mathcal{O}(Time_\Gamma(n))$.

Proof. Let $\Gamma = (V, U, G, \mathcal{N}, \alpha, \beta, x_1, x_n)$ be a NEP with the underlying graph $G = (X_G, E_G)$ and $X_G = \{x_1, x_2, \ldots, x_n\}$ for some $n \geq 1$; $x_1 \equiv \underline{In}$ and $x_n \equiv \underline{Halt}$. We construct the NEP $\Gamma = (V', U', G', \mathcal{N}', \alpha', \beta', x_{start}, x_n^s)$; $x_{start} \equiv \underline{In}$ and $x_n^s \equiv \underline{Halt}$, where

$$V' = V, \qquad U' = U \cup T,$$
$$T = \{t_i^l, t_i^r, t_i^{l'}, t_i^{r'}, t_i^{l''}, t_i^{r''} \mid 1 \leq i \leq n\}$$

Note that the underlying graph G' is a complete graph. First, we add the following nodes to G':

- <u>node x_{start}</u>:

$$M = \begin{cases} \{\varepsilon \to t_1^{l''}\}, \text{ if } \alpha(x_1) \neq l \\ \{\varepsilon \to t_1^{r''}\}, \text{ if } \alpha(x_1) = l \end{cases},$$
$$PI = \emptyset, \qquad FI = T,$$
$$PO = \emptyset, \qquad FO = \emptyset,$$
$$\alpha = \begin{cases} l, \text{ if } \alpha(x_1) \neq l \\ r, \text{ if } \alpha(x_1) = l \end{cases}, \qquad \beta = (w).$$

- nodes x_i^s, $1 \le i \le n$ (they actually simulate the work of x_i in Γ):
$$\begin{array}{ll} M = M(x_i), & \\ PI = PI(x_i), & FI = FI(x_i) \cup T \setminus \{t_i^l, t_i^r\}, \\ PO = PO(x_i), & FO = FO(x_i), \\ \alpha = \alpha(x_i), & \beta = \beta(x_i). \end{array}$$

For each node x_i, $1 \le i \le n$ in Γ we add a subnetwork to Γ' according to the subsequent cases:

Case 1. If $\alpha(x_i) = l$, the subnetwork is defined as follows (these nodes are used for preparing the string in the aim of processing them in the nodes x_i^s):

- nodes x_i^{Ins}, $1 \le i \le n$:
$$\begin{array}{ll} M = \{\varepsilon \to t_i^{r''}\}, & \\ PI = \{t_i^{l'}\}, & FI = \emptyset, \\ PO = \{t_i^{r''}\}, & FO = \emptyset, \\ \alpha = r, & \beta = (w). \end{array}$$

- nodes x_i^{Del}, $1 \le i \le n$:
$$\begin{array}{ll} M = \{t_i^{l'} \to \varepsilon\}, & \\ PI = \{t_i^{r''}\}, & FI = \emptyset, \\ PO = \emptyset, & FO = \emptyset, \\ \alpha = l, & \beta = (w). \end{array}$$

- nodes x_i^{Sub}, $1 \le i \le n$:
$$\begin{array}{ll} M = \{t_i^r \to t_j^{r'} \mid \{x_i, x_j\} \in \Gamma\} \cup & \\ \{t_i^{r'} \to t_i^r\} \cup \{t_i^{r''} \to t_i^r\}, & \\ PI = \{t_i^r, t_i^{r'}, t_i^{r''}\}, & FI = \{t_i^{l'}\}, \\ PO = \emptyset, & FO = \emptyset, \\ \alpha = *, & \beta = (w). \end{array}$$

Case 2. If $\alpha(x_i) = r$, the subnetwork is analogous to the *Case 1* with the characters l and r interchanged.

Case 3. If $\alpha(x_i) = *$, the subnetwork is defined as follows (the role of these nodes is the same as above, namely to prepare the strings for being processed in the nodes x_i^s):

- nodes x_i^{Sub}, $1 \le i \le n$:
$$\begin{array}{ll} M = \{t_i^r \to t_j^{r'} \mid \{x_i, x_j\} \in \Gamma\} \cup & \\ \{t_i^l \to t_j^{l'} \mid \{x_i, x_j\} \in \Gamma\} \cup & \\ \{t_i^{r'} \to t_i^r\} \cup \{t_i^{l'} \to t_i^l\} \cup \{t_i^{l''} \to t_i^l\}, & \\ PI = \{t_i^l, t_i^r, t_i^{l'}, t_i^{r'}, t_i^{l''}\}, & FI = \emptyset, \\ PO = \emptyset, & FO = \emptyset, \\ \alpha = *, & \beta = (w). \end{array}$$

Let w be the input string in Γ. In the input node x_{start}, the character $t_1^{l''}$ is inserted at the beginning of the string if $\alpha(x_1) \in \{r, *\}$, or the character $t_1^{r''}$ is inserted at the end of the string, provided that $\alpha(x_1) = l$. Next, the string enters x_1^{Sub} where the character is replaced with t_1^l and t_1^r, respectively. Then, the string can only enter x_1^s and the simulation starts. Note that the same evolutionary rules applicable in $x_1 \in \Gamma$ are also possible in x_1^s since the special character $t_1^{l''}$ or $t_1^{r''}$ is set up in a way that it does not block the computation of nodes with $\alpha = r$ and $\alpha = l$, respectively. Inductively, we may assume that a string of the form $t_i^l w$ or wt_i^r lies in the node $x_i^s \in \Gamma'$ if and only if the string w lies in the node $x_i \in \Gamma$.

Let w be transformed into w' in the node x_i and sent to the connected nodes to x_i in Γ. Then, a string $t_i^l w'$ or a string wt_i^r is produced in the node x_i^s and sent to the node x_i^{Sub}. Let us analyze the case of a string $t_i^l w'$. The process is analogous for the other string. In x_i^{Sub}, the character t_i^l is replaced with the symbol $t_j^{l'}$, assuming that $\{x_i, x_j\} \in \Gamma$, which ensures the new string can only be accepted by subnetworks j corresponding to nodes x_j connected to x_i in the original network Γ. From here, the process differs in accordance with the value α of the connected node x_j.

- If $\alpha(x_j) = l$, the string can only enter x_j^{Ins} where the symbol $t_j^{r''}$ is appended to it. The new string, $t_j^{l'} w' t_j^{r''}$, continues through x_j^{Del} where $t_j^{l'}$ is removed and x_j^{Sub} where $t_j^{r''}$ is replaced with t_j^r, allowing it to enter the node x_j^s. Since the character t_j^r is at the end of the string, it does not interfere with the application of evolutionary rules at the left of the string;
- If $\alpha(x_j) = r$ or $\alpha(x_j) = *$, the string directly enters x_j^{Sub} and the symbol $t_j^{l'}$ is replaced with t_j^l. Then, the string enters x_j^s. As one can see, the communication step in Γ has been simulated by a constant number of (evolution and communication) steps in Γ'. A new evolutionary step in Γ is now simulated. It follows that $L(\Gamma) = L(\Gamma')$. Furthermore, the number of steps in Γ' for simulating an evolutionary step followed by a communication one in Γ is constant; hence, $Time_{\Gamma'}(n) \in \mathcal{O}(Time_{\Gamma}(n))$ holds.

□

4. Simulating Any NEP with a Star NEP

Theorem 2. *Given an arbitrary NEP Γ, there exists a star NEP Γ' such that the following two conditions are satisfied:*

1. $L(\Gamma) = L(\Gamma')$;
2. $Time_{\Gamma'}(n) \in \mathcal{O}(Time_{\Gamma}(n))$.

Proof. Let $\Gamma = (V, U, G, \mathcal{N}, \alpha, \beta, x_1, x_n)$ be a NEP with the underlying graph $G = (X_G, E_G)$ and $X_G = \{x_1, x_2, \ldots, x_n\}$ for some $n \geq 1$; $x_1 \equiv \underline{In}$ and $x_n \equiv \underline{Halt}$. We construct the NEP $\Gamma = (V', U', G', \mathcal{N}', \alpha', \beta', x_{start}, x_n^s)$; $x_{start} \equiv \underline{In}$ and $x_n^s \equiv \underline{Halt}$, where

$$V' = V, \qquad U' = U \cup T,$$
$$T = \{t_i^l, t_i^r, t_i^{l'}, t_i^{r'}, t_i^{l''}, t_i^{r''}, t_i^{l'''}, t_i^{r'''} \mid 1 \leq i \leq n\}$$

The *star* network uses the definitions illustrated above for the *complete* network, with the following modifications:

We add a new node *Star* to the subnetwork which acts as the center of the *star* network.

- $\underline{\text{node } Star}$:
$$\begin{aligned} &M = \{t_i^l \to t_j^{l'} \mid \{x_i, x_j\} \in \Gamma\} \cup \{t_i^r \to t_j^{r'} \mid \{x_i, x_j\} \in \Gamma\} \cup \\ &\{t_i^{l'''} \to t_i^l\} \cup \{t_i^{r'''} \to t_i^r\}, \\ &PI = \emptyset, \qquad FI = \emptyset, \\ &PO = \emptyset, \qquad FO = \emptyset, \\ &\alpha = *, \qquad \beta = (w). \end{aligned}$$

The nodes $x_i^{Sub}, 1 \leq i \leq n$ are modified as follows:

Case 1. If $\alpha(x_i) = l$:

- $\underline{\text{nodes } x_i^{Sub}, 1 \leq i \leq n}$:
$$\begin{aligned} &M = \{t_i^{r'} \to t_i^{r'''}\} \cup \{t_i^{r''} \to t_i^{r'''}\}, \\ &PI = \{t_i^{r'}, t_i^{r''}\}, \qquad FI = \{t_i^{l'}\}, \\ &PO = \emptyset, \qquad FO = \emptyset, \\ &\alpha = *, \qquad \beta = (w). \end{aligned}$$

Case 2. If $\alpha(x_i) = r$, the nodes x_i^{Sub} are analogous to the *case 1* with the characters l and r interchanged.

Case 3. If $\alpha(x_i) = *$, the nodes $x_i^{Sub}, 1 \leq i \leq n$ are defined in the following way:

- $\underline{\text{nodes } x_i^{Sub}, 1 \leq i \leq n}$:
$$\begin{aligned} &M = \{t_i^{r'} \to t_i^{r'''}\} \cup \{t_i^{l'} \to t_i^{l'''}\} \cup \\ &\{t_i^{l''} \to t_i^{l'''}\}, \\ &PI = \{t_i^{l'}, t_i^{r'}, t_i^{l''}\}, \qquad FI = \emptyset, \\ &PO = \emptyset, \qquad FO = \emptyset, \\ &\alpha = *, \qquad \beta = (w). \end{aligned}$$

Let w be the input string in Γ. In the input node x_{start}, the character $t_1^{l''}$ is inserted in the left-hand side of the string if $\alpha(x_1) \in \{r, *\}$, or the character $t_1^{r''}$ is inserted at the end of the string provided that $\alpha(x_1) = l$. Next, the string enters *Star* where no rule can be applied. From *Star*, it can only enter x_1^{Sub} where the character is replaced with $t_1^{l'''}$ and $t_1^{r'''}$, respectively. The new string returns to *Star* where $t_1^{l'''}$ and $t_1^{r'''}$ are changed to t_1^l and t_1^r. Then, the string can only enter x_1^s and the simulation starts. Note that the same evolutionary rules applicable in $x_1 \in \Gamma$ are also possible in x_1^s since the special character $t_1^{l''}$ or $t_1^{r''}$ is set up in a way that it does not block the computation of nodes with $\alpha = r$ and $\alpha = l$, respectively. Inductively, we may assume that a string of the form $t_i^l w$ or wt_i^r lies in the node $x_i^s \in \Gamma'$ if and only if the string w lies in the node $x_i \in \Gamma$.

Let w be transformed into w' in the node x_i and sent to the connected nodes to x_i in Γ. Then, a string $t_i^l w'$ or a string $w' t_i^r$ is produced in the node x_i^s and sent to the node *Star*. Let us analyze the case of a string $t_i^l w'$. The process is analogous for the other string. In *Star*, the character t_i^l is replaced with the symbol $t_j^{l'}$, granted that $\{x_i, x_j\} \in \Gamma$, which ensures the new string can only be accepted by subnetworks j corresponding to nodes x_j connected to x_i in the original network Γ. From here, the process is similar to the one described in the previous proof.

- If $\alpha(x_j) = l$, the string can only enter x_j^{Ins} where the symbol $t_j^{r''}$ is attached at the end of it. The new string, $t_j^{l'} w' t_j^{r''}$, continues through x_j^{Del} where $t_j^{l'}$ is removed and x_j^{Sub} where $t_j^{r''}$ is replaced with $t_j^{r'''}$. Then, $t_j^{r'''}$ is switched with t_j^r in *Star*, allowing it to enter the node x_j^s. Since the character t_j^r is at the end of the string, it does not interfere with the application of evolutionary rules at the left of the string;
- If $\alpha(x_j) = r$ or $\alpha(x_j) = *$, the string directly enters x_j^{Sub} and the symbol $t_j^{l'}$ is replaced with $t_j^{l'''}$. Then, the string enters x_j^s after having $t_j^{l'''}$ changed to t_j^l in *Star*. As in the previous construction, the communication step in Γ has been simulated by a constant number of (evolution and communication) steps in Γ', and a new evolutionary step in Γ is going to be simulated. We conclude that the two networks accept the same language.

The explanations above allow us to infer that any step in Γ is simulated by a constant number of steps in Γ'; hence, $Time_{\Gamma'}(n) \in \mathcal{O}(Time_{\Gamma}(n))$ holds. □

5. Simulating Any NEP with a Grid NEP

Theorem 3. *Given an arbitrary NEP Γ there exists a grid NEP Γ' such that the following two conditions are satisfied:*

1. $L(\Gamma) = L(\Gamma')$;
2. $Time_{\Gamma'}(n) \in \mathcal{O}(Time_{\Gamma}(n))$.

Proof. Let $\Gamma = (V, U, G, \mathcal{N}, \alpha, \beta, x_1, x_n)$ be a NEP with the underlying graph $G = (X_G, E_G)$ and $X_G = \{x_1, x_2, \ldots, x_n\}$ for some $n \geq 1$; $x_1 \equiv \underline{In}$ and $x_n \equiv \underline{Halt}$. We construct the NEP $\Gamma = (V', U', G', \mathcal{N}', \alpha', \beta', x_{start}, x_n^s)$; $x_{start} \equiv \underline{In}$ and $x_n^s \equiv \underline{Halt}$, where

$$V' = V, \qquad U' = U \cup T,$$
$$T = \{t_i^l, t_i^r, t_i^{l'}, t_i^{r'} \mid 1 \leq i \leq n\}$$

First, we add the following nodes to Γ':

- $\underline{\text{node } x_{start}}$:
$$M = \begin{cases} \{\varepsilon \to t_1^l\}, \text{ if } \alpha(x_1) \neq l \\ \{\varepsilon \to t_1^r\}, \text{ if } \alpha(x_1) = l \end{cases}, \qquad PI = \emptyset, \quad FI = T, \quad PO = \emptyset, \quad FO = \emptyset,$$
$$\alpha = \begin{cases} l, \text{ if } \alpha(x_1) \neq l \\ r, \text{ if } \alpha(x_1) = l \end{cases}, \qquad \beta = (w).$$

- $\underline{\text{nodes } x_i^s,\ 1 \le i \le n:}$ $\begin{array}{ll} M = M(x_i), & \\ PI = PI(x_i), & FI = FI(x_i) \cup T \setminus \{t_i^l, t_i^r\}, \\ PO = PO(x_i), & FO = FO(x_i), \\ \alpha = \alpha(x_i), & \beta = \beta(x_i). \end{array}$

For each node x_i, $1 \le i \le n$ in Γ we add a subnetwork to Γ' according to the subsequent cases:

Case 1. If $\alpha(x_i) = l$, the subnetwork is defined as follows:

- $\underline{\text{nodes } x_i^{Ins},\ 1 \le i \le n:}$ $\begin{array}{ll} M = \{\varepsilon \to t_i^{r\prime}\}, & \\ PI = \emptyset, & FI = T, \\ PO = \{t_i^{r\prime}\}, & FO = \emptyset, \\ \alpha = r, & \beta = (w). \end{array}$

- $\underline{\text{nodes } x_i^{Del},\ 1 \le i \le n:}$ $\begin{array}{ll} M = \{t_i^{l\prime} \to \varepsilon\}, & \\ PI = \{t_i^{l\prime}\}, & FI = \emptyset, \\ PO = \emptyset, & FO = \emptyset, \\ \alpha = l, & \beta = (w). \end{array}$

- $\underline{\text{nodes } x_i^{Sub},\ 1 \le i \le n:}$ $\begin{array}{ll} M = \{t_i^r \to t_j^{r\prime} \mid \{x_i, x_j\} \in \Gamma\} \cup \{t_i^{r\prime} \to t_i^r\} \cup & \\ \{t_i^{r\prime\prime} \to t_i^r\}, & \\ PI = T, & FI = \emptyset, \\ PO = \emptyset, & FO = \emptyset, \\ \alpha = *, & \beta = (w). \end{array}$

Case 2. If $\alpha(x_i) = r$, the subnetwork is analogous to the *case 1* with the symbols l and r interchanged.

Case 3. If $\alpha(x_i) = *$, the subnetwork is defined as follows:

- $\underline{\text{nodes } x_i^{Sub},\ 1 \le i \le n:}$ $\begin{array}{ll} M = \{t_i^r \to t_j^{r\prime} \mid \{x_i, x_j\} \in \Gamma\} \cup & \\ \{t_i^l \to t_j^{l\prime} \mid \{x_i, x_j\} \in \Gamma\} \cup & \\ \{t_i^{l\prime} \to t_i^l\} \cup \{t_i^{r\prime} \to t_i^r\}, & \\ PI = T, & FI = \emptyset, \\ PO = \emptyset, & FO = \emptyset, \\ \alpha = *, & \beta = (w). \end{array}$

Lastly, we add a set of dummy nodes to complete the grid topology with the specifications below:

- $\underline{\text{nodes } D_i,\ 1 \le i \le 2n \wedge \alpha(x_i) = *:}$ $\begin{array}{ll} M = \emptyset, & \\ PI = \emptyset, & FI = U', \\ PO = \emptyset, & FO = \emptyset, \\ \alpha = *, & \beta = (w). \end{array}$

- $\underline{\text{nodes } D:}$ $\begin{array}{ll} M = \emptyset, & \\ PI = \emptyset, & FI = \{t_i^l, t_i^r \mid 1 \le i \le n\}, \\ PO = \emptyset, & FO = \emptyset, \\ \alpha = *, & \beta = (w). \end{array}$

The grid network is set up in the following way.

- The node x_{start} is in the top left corner. The first column is composed by it followed by the node x_1^s corresponding to the input node $x_1 \in \Gamma$ and the remaining nodes x_i^s arranged in any order;
- The second column is composed by a dummy node D and the nodes x_i^{Sub}. Each node x_i^{Sub} is connected to the node x_i^s through the left edge;

- The third column is composed by a dummy node D and the nodes x_i^{Del}. Each node x_i^{Del} is connected to the node x_i^{Sub} through the left edge. In the case of $\alpha = *$, a node D_i is used instead of a node x_i^{Del};
- The fourth column is composed by a dummy node D and the nodes x_i^{Ins}. Each node x_i^{Ins} is connected to the node x_i^{Del} through the left edge. In the case of $\alpha = *$, a node D_i is used instead of a node x_i^{Ins};
- The fifth column is composed by nodes D.

Let w be the input string in Γ. In the input node x_{start}, the character t_1^l is inserted in the beginning of the string if $\alpha(x_1) \in \{r, *\}$, or the character t_1^r is inserted at the end of the string, if $\alpha(x_1 \in \Gamma) = l$. Then, the string can only enter x_1^s and the simulation starts. Note that the same evolutionary rules applicable in $x_1 \in \Gamma$ are also possible in x_1^s since the special character t_1^l or t_1^r is set up in a way that it does not block the computation of nodes with $\alpha = r$ and $\alpha = l$, respectively. Inductively, we may assume that a string of the form $t_i^l w$ or wt_i^r lies in the node $x_i^s \in \Gamma'$ if and only if the string w lies in the node $x_i \in \Gamma$.

Let w be transformed into w' in the node x_i and sent to the connected nodes to x_i in Γ. Then, a string $t_i^l w'$ or a string wt_i^r is produced in the node x_i^s and sent to the connected node x_i^{Sub}. In this node, the symbols t_i^l and t_i^r are replaced with $t_j^{l'}$ and $t_j^{r'}$, respectively, granted that $\{x_i, x_j\} \in \Gamma$. Then, the string continues through the second column of x_i^{Sub} nodes until it ultimately enters the node x_j^{Sub}. Note that even if the string passes through the other nodes $x_k^{Sub} \mid k \neq j$, no rule can applied so the string remains unchanged until it gets to the desired node. Next, the computation can be continued in one of the following ways:

- If $\alpha(x_j) = l$, no rule can be applied in x_j^{Sub} and the string enters x_j^{Del}. In that node, the symbol $t_j^{l'}$ is removed. Next, since it does not contain any character $t \in T$, the string can only enter the node x_j^{Ins} where a character $t_j^{r'}$ is attached to the end. Then, the string continues through the fifth column of dummy nodes D and it ultimately returns to x_j^{Sub} where $t_j^{r'}$ is replaced with t_j^r, allowing it to enter the node x_j^s;
- If $\alpha(x_j) = r$ or $\alpha(x_j) = *$, the string directly enters x_j^{Sub} and the symbol $t_j^{l'}$ is replaced with t_j^l. Then, the word enters x_j^s. As in the previous proofs, we conclude that $L(\Gamma) = L(\Gamma')$, as well as $Time_{\Gamma'}(n) \in \mathcal{O}(Time_{\Gamma}(n))$.

□

6. Conclusions and Further Work

We have proposed three constructions for simulating an arbitrary NEP by a NEP having an underlying structure that is a complete graph, a star graph, and a two-dimensional grid, respectively. All these simulations are time efficient in the sense that every computational step in the given network is simulated by a constant number of computational steps in the constructed network.

In our view, it would be of interest whether or not similar results are valid for other variants of NEPs, such as polarized NEPs or NEPs with filtered connections as well as for variants of networks of splicing processors.

Author Contributions: Conceptualization, V.M.; methodology, V.M. and J.Á.S.M.; validation, V.M. and J.Á.S.M.; formal analysis, J.Á.S.M.; investigation, V.M. and J.Á.S.M.; writing—original draft preparation, J.Á.S.M.; writing—review and editing, V.M. and J.Á.S.M.; supervision, V.M.; funding acquisition, V.M. Both authors have read and agreed to the published version of the manuscript.

Funding: This work was partially supported by a grant of the Romanian Ministry of Education and Research, CCCDI-UEFISCDI, Project No. PN-III-P2-2.1-PED-2019-2391, within PNCDI III.

Conflicts of Interest: The authors declare no conflict of interest.

References

1. Castellanos, J.; Martín-Vide, C.; Mitrana, V.; Sempere, J.M. Networks of evolutionary processors. *Acta Inform.* **2003**, *39*, 517–529. [CrossRef]
2. Csuhaj-Varjú, E.; Martín-Vide, C.; Mitrana, V. Hybrid networks of evolutionary processors are computationally complete. *Acta Inform.* **2005**, *41*, 257–272. [CrossRef]
3. Manea, F.; Margenstern, M.; Mitrana, V.; Pérez-Jiménez, M.J. A new characterization of NP, P, and PSPACE with accepting hybrid networks of evolutionary processors. *Theory Comput. Syst.* **2010**, *46*, 174–192. [CrossRef]
4. Manea, F.; Mitrana, V. All **NP**-problems can be solved in polynomial time by accepting hybrid networks of evolutionary processors of constant size. *Inf. Process. Lett.* **2007**, *103*, 112–118. [CrossRef]
5. Bottoni, P.; Labella, A.; Manea, F.; Mitrana, V.; Petre, I.; Sempere, J.M. Complexity-preserving simulations among three variants of accepting networks of evolutionary processors. *Nat. Comput.* **2011**, *10*, 429–445. [CrossRef]
6. Alarcón, P.; Arroyo, F.; Mitrana, V. Networks of polarized evolutionary processors. *Inform. Sci.* **2014**, *265*, 189–197. [CrossRef]
7. Drăgoi, C.; Manea, F.; Mitrana, V. Accepting networks of evolutionary processors with filtered connections. *J. Univ. Comput. Sci.* **2007**, *13*, 1598–1614.
8. Navarrete, C.; Cruz, M.; Rey, E.; Ortega, A.; Rojas, J. Parallel simulation of NEPs on clusters. In Proceedings of the International Conferences on Web Intelligence and Intelligent Agent Technology, Lyon, France, 22–27 August 2011; Volume 3, pp. 171–174.
9. Gómez, S.; Ortega, A.; Orgaz, P. Distributed simulation of NEPs based nn-demand cloud elastic computation. *Adv. Comput. Intell.* **2015**, *9094*, 40–54.
10. Gómez, S.; Ordozgoiti, B.; Mozo, A. NPEPE: Massive natural computing engine for optimally solving NP-complete problems in Big Data scenarios. *New Trends Databases Inf. Syst.* **2015**, *539*, 207–217.
11. Gómez, S.; Mitrana, V.; Păun, M.; Vararuk, S. High performance and scalable simulations of a bio-inspired computational model. In Proceedings of the International Conference on High Performance Computing & Simulation, Dublin, Ireland, 15–19 July 2019; pp. 543–550.
12. Rozenberg, G.; Salomaa, A. (Eds.) *Handbook of Formal Languages*; Springer: Berlin, Germany, 1998.
13. Manea, F.; Martín-Vide, C.; Mitrana, V. On the size complexity of universal accepting hybrid networks of evolutionary processors. *Math. Struct. Comput. Sci.* **2007**, *17*, 753–771. [CrossRef]
14. Loos, R.; Manea, F.; Mitrana, V. Small universal accepting hybrid networks of evolutionary processors. *Acta Inform.* **2010**, *47*, 133–146. [CrossRef]

axioms

MDPI

Article

Interdimensionality

Karl Svozil

Institute for Theoretical Physics, TU Wien, Wiedner Hauptstrasse 8-10/136, 1040 Vienna, Austria; svozil@tuwien.ac.at

Abstract: In this speculative analysis, interdimensionality is introduced as the (co)existence of universes embedded into larger ones. These interdimensional universes may be isolated or intertwined, suggesting a variety of interdimensional intrinsic phenomena that can only be understood in terms of the outer, extrinsic reality.

Keywords: intrinsic perception; Hausdorff dimension; fractal

MSC: 00A73; 28A78; 28A80; 37C45; 54F35; 54F45; 83E15

1. A Caveat: Speculation and Progress

Rule inference is the process of hypothesizing a general rule or "law" from examples or "phenomena" [1,2]. The halting problem is the task to determine, given an arbitrary computer program and an input, whether the program will eventually halt or continue to run forever. This has been proven to be unsolvable in general. As the former rule inference problem can be reduced to the latter halting problem, it is provable unsolvable in general. This constraint on induction has been coped with by the philosophy of science in a variety of ways:

Popper suggested that, instead of induction and verification, which appears to be a hopeless endeavor, falsification might be a good demarcation criterion between science on the one hand, and on the other hand ideology, sophisms, or, in a more frugal term, bullshit [3]. Lakatos responded by criticizing that, due to side assumptions and a vast 'protective belt' of auxiliary hypotheses, in many practical circumstances, falsification fails. As a result, contemporaries can seldom predict what might turn out to become a progressive versus a degenerative research program [4].

Kuhn observed that science may be characterized by brief iconoclastic periods of revolution, followed by longer conformist periods of consolidation [5]. Feyerabend even challenged methodology as mythology and ideology akin to religious dogmas, and suggested keeping science wide open and performing an "exhaustive search" of ideas by allowing "anything" to enter the scientific debate, thereby, imposing little methodological restrictions [6]; he also recommended a formal separation between state and science, and lay judges for the evaluation of success [7] and the allocation of scientific funding.

In any case, there seems to be no convergence of conceptual progression. Taking gravity and celestial motion, for example: the Ptolemaic system was expressed in terms of geometry. It was superseded by the Copernican revolution that later became based on Newtonian gravitational forces. Later on, Newtonian gravity was replaced by the curved geometry of space–time of Einstein's theory of general relativity. By analogy, it appears highly likely that our contemporaries would view any model superseding the present canon as utterly speculative, if not outright nonsense.

Such a historic perspective leads to greater liberty and openness of ideas, and yet this creativity needs to be guided and stimulated by empirical findings and attempts to falsify consequences and claims. This amounts to an amalgam of the aforementioned ideas brought forward in the philosophy of science, resulting in a sort of pragmatism that is

Citation: Svozil, K. Interdimensionality. *Axioms* **2021**, *10*, 300. https://doi.org/10.3390/axioms10040300

Academic Editor: Palle E. T. Jorgensen

Received: 19 October 2021
Accepted: 9 November 2021
Published: 12 November 2021

well balanced between wild fantasy and empirical grounding. Exactly how much of those ingredients are in order may greatly depend on the temperament and character of the individual researcher.

We, therefore, present the following considerations with a caveat to the reader, as it trespasses far beyond any empirically verifiable physics of our time; and yet some aspects of it might indicate the way to fruitful avenues of scientific modeling. We hope that the following speculations are not too weird for the realistic, critical, and sober mind. At best this could be seen as a vision of things to come.

2. Definition

Interdimensionality, or, by another naming, dimensional shadowing [8]—the "emulation" of a lowerdimensional configuration space by a fractal subset of a higherdimensional manifold—is the (co)existence and (co)habitation of parts or fragments of an "outer" space of a "higher" extrinsic Hausdorff dimension [9] by some "inner" subspace entity that has a "lower" or equal intrinsic Hausdorff dimension. One may imagine such a situation as a fractal of Hausdorff dimension d embedded in a continuum, such as the Hilbert space $\mathbb{R}^n$ or $\mathbb{C}^n$, with $d \leq n$. Therefore, pointedly speaking, we might exist on a sort of Cantor set or Menger sponge-like structure—fractals obtained by self-similar elimination of proper parts—of (almost) an integer Hausdorff dimension, which is part of a high-dimensional super-verse.

Formally, the Hausdorff dimension d of a set $A \in \mathbb{R}^n$, defined via the d-dimensional Hausdorff measure, is based on its "umklapp" property—the sudden change from measure value zero to infinity if the dimension parameter is taken higher or lower than a unique value—as follows. Suppose $\cup_i F_i$ covers A, and suppose further that there exists a limit in which all individual constituents F_i of this covering become infinitesimal in diameter. Then, the Hausdorff measure μ_d, and a unique dimensional parameter d called the Hausdorff dimension is

$$\mu_\delta(A) = \lim_{\epsilon \to 0+} \inf_{\{F_i\}} \left\{ \sum_i (\text{diam } F_i)^\delta \,\middle|\, \delta \in \mathbb{R},\ \delta > 0,\ \cup_i F_i \supset A,\ (\text{diam } F_i) \leq \epsilon \right\}, \tag{1}$$

where the infimum is over all countable ϵ-covers $\{F_i\}$ of A; with the dimension d as an "umklapp" parameter of

$$\mu_\delta(A) = \begin{cases} 0 & \text{if } \delta > d, \\ \infty & \text{if } \delta < d. \end{cases} \tag{2}$$

That is, the Hausdorff dimension d is the unique dimensional parameter at which the measure μ_δ as a function of the dimensional parameter value δ smaller or larger than d is infinite or vanishes, respectively. Note that the diameter "diam" presupposes the notion of a distance defined via a metric. For self-similar fractal sets, the capacity dimension c is defined by

$$c = \lim_{\epsilon \to 0+} \log[n(\epsilon)] / \log\left(\epsilon^{-1}\right), \tag{3}$$

where $n(\epsilon)$ is the number of segments of length ϵ, equals the Hausdorff dimension d.

An example of a set of integer dimension m embedded into an outer space $\mathbb{R}^n$ with $n > m$ is the set whose (contravariant) coordinates with respect to some (covariant) basis $\mathbb{R}^n$ is given by

$$\left\{ \left(x_1, x_2, \dots, x_m, r_1(x_1, x_2, \dots, x_m), \dots, r_{n-m}(x_1, x_2, \dots, x_m) \right)^\intercal \middle| x_i, r_j \in \mathbb{R} \right\}, \tag{4}$$

where $r_i(x_1, x_2, \dots, x_m)$, $1 \leq i \leq n - m$ are some total, possibly constant or random, choice functions.

For most practical operational purposes [10,11] the intrinsic perception of the dimensionality of such shadowed, interdimensional object might effectively remain that of a "solid continuum" of that intrinsic (Hausdorff) dimension. It may not be too unreasonable to compare this to the common notion of "emptiness of space in-between point particles" constituting solid physical objects, or the "perceived continuous motion" from individual still frames [12,13].

There are some findings consistent such speculations: For instance, associated with every integer-dimensional regular rectifiable m-dimensional fractal embedded in $\mathbb{R}^n$, there exists a locally defined tangential m–dimensional vector subspace of $\mathbb{R}^n$ [9,14]. Even for non-integer-dimensional fractals, integer-dimensional tangent spaces may be "good" approximations for all practical physical purposes.

Further examples for cohabitation of continua that need not involve fractals are paradoxical decompositions, such as Vitali's partition of the unit interval and the decomposition of the sphere by Hausdorff [15]. If we relax the definition of dimension, we may also speak of (dense) "scattered" point sets "inhabiting" the continuum. The variations may be manyfold; for instance, one may consider partitions or intertwined subsets of continua. One may not even deal with extrinsic continua but with general sets that allow some form of intrinsic embedding.

Let us finally review two almost trivial examples of an arbitrary number of one-dimensional subspaces of $\mathbb{R}^2$, as schematically depicted in Figure 1. The first one is a collection of parallel lines. The second one is a star-shaped configuration intertwining in the origin, spanned by respective mutually distinct unit vectors. In the latter case, the only way for "flatlanders" [16] living on different subspaces to communicate with each other is through a single point—the origin.

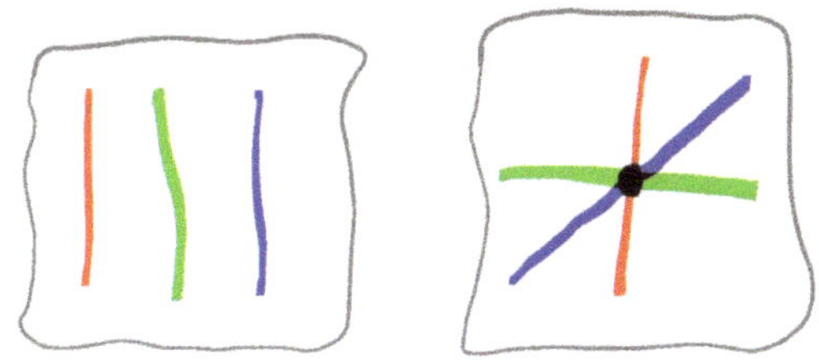

(a) (b)

Figure 1. Schematic drawing of interdimensional configurations that are (**a**) isolated or (**b**) intertwine, as seen from some outer, embedding space.

In general, fractals need not be regular and rectifiable and of integer dimension. Rather they may be "cloud-like shapes", with "scattered" holes and gaps. Those gaps will not be perceived intrinsically. Indeed, one may speculate that this situation gives rise to a metric that essentially mimics curvature [17].

Fractal theory has inspired and evolved into many innovative, useful, and interesting applications, especially in new materials and nanostructures. Such important developments can lead us to new views of, and physical means related to, dimensionality [18,19].

As the aim is the provision of a very general analysis that is unconstrained by the technicalities of specific models, no concrete theory is discussed. Nevertheless, it might be not too far-fetched to briefly mention some potential connections between interdimensionality and various paradigms in modern particle physics and cosmology. Some of these involve the description of a volume of space as conceptualized by holographic principles, such as the AdS/CFT correspondence related to D-branes in string theory, or the ekpyrotic models relying on string theory, branes, and extra "hidden" dimensions. Other scenarios in the context of the theory of general relativity involve traversable wormholes (aka Einstein–Rosen bridges) linking disparate points in spacetime.

3. Disjoint and Intertwining Shadows

To proceed to interdimensional motion, we need to consider intertwining areas of interdimensionality. The simplest nontrivial case is the one schematically depicted in

Figure 1b in which all universes share a single point of communication. Of greater interest might be a situation in which an entire region of space is shared. One might think also of a "small" fraction of a universe "traversing" another universe, such that, compared to the overall extension of these universes, this common share appears like the tip of an iceberg.

4. Interdimensional Motion

Interdimensional motion is the motion of some "inner" intrinsic subspace in the "outer", extrinsic space. If two inner spaces are involved, it may happen that certain limits of motion, such as continuity or maximal speed, that are valid in one subspace, can be breached and overcome by another subspace. In what follows, some scenarios will be discussed. We shall adopt the following notation: inner "intrinsic" subspaces will be denoted by M and N.

Let us discuss this by considering a simple example of a rotating point, as schematically drawn in Figure 2a. From the point of view of M the rotation in N is observed as periodic (dis)appearances of some object rotating in M.

Another "wormhole"-like scenario schematically drawn in Figure 2b is a "bend" or "curved" (relative to the exterior "outer" continuum) reference frame M that is intermittantly accessed from N. Suppose that the propagation speed limit for motion is the same $c_M = c_N$ in both frames. Then, the object appears to be traveling with a velocity greater than this limit velocity in M because of the "shortcut" access through N.

Still another scenario schematically drawn in Figure 2c is one in which N allows for faster that M–light motion—that is, $c_M \ll c_N$—and this property is used to access regions in M through motion in N that appear space-like separated in M's frame of reference.

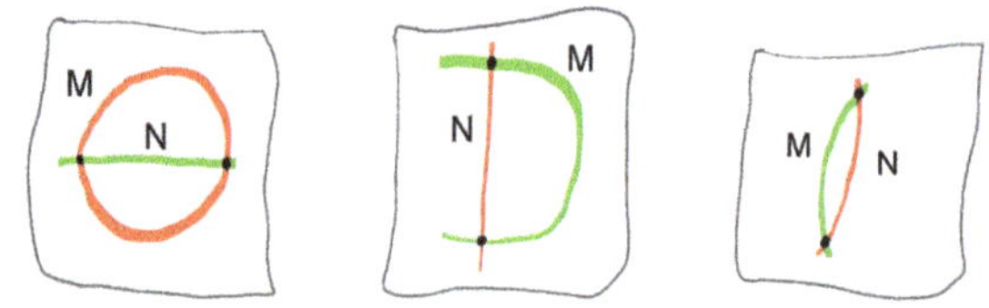

(a) (b) (c)

Figure 2. Schematic drawing of worldlines of interdimensional motion, as seen from the outer, embedding space: (**a**) periodic, (**b**) shortcut, and (**c**) coevolution.

4.1. Interdimensional Chronology Protection

In these and similar situations, no issues with respect inconsistent evolution, in particular, time paradoxes, arise. As whatever relative space–time reference frames are operationally constructed [20] in M and N, the "outer" extrinsic space, in which both M and N are embedded, regulates the phenomenology.

Indeed, from an extrinsic, "God's eye view" of the outer space there is no consistency issue because the evolution seen from this "global" comprehensive perspective never yields or allows inconsistent phenomena. Concerns raised by intrinsic space–time frames generated with the means available in M and N are merely epistemic, and the means are relative to the devices and conventions (such as for synchronizing clocks) available to the inhabitants of M and N.

This results in an interdimensional scheme of chronology protection based on the epistemic relativity of reference frames. At the same time, from an "outer" (i.e., ontological) point of view, those frames are "bundled together" through the coembedding and cohabitation of some outer space.

There are similarities between the consistency of observable phenomena regarding the higher-dimensional bulk space and the consistent histories approach to the Many Worlds models [21]. Both involve multiple "merging" paths.

4.2. Examples of Dimensional Relativity

The following examples closely follow the scenarios schematically depicted in Figure 2b,c. They have some similarities to ballistic missiles that avoid the limitations of velocity from atmospheric drag (friction) by leaving and re-entering Earth's atmosphere, or are analogs of supercavitation—the formation of vapor bubbles in a liquid caused by flow around an object, allowing minimal friction movement inside liquids at nearly the speed of sound.

The first example, depicted in Figure 3, shows an interdimensional dive into a dimension that allows higher velocities, or rather traversals of space per time, in M through "jump" into another dimension N, thereby, creating a shortcut from two space–time points A to B. This is different from breaking the intradimensional warp barrier by hyper-fast solitons in Einstein–Maxwell-plasma theory [22] as it employs dimensional capacities that are not bound by intradimensional motion.

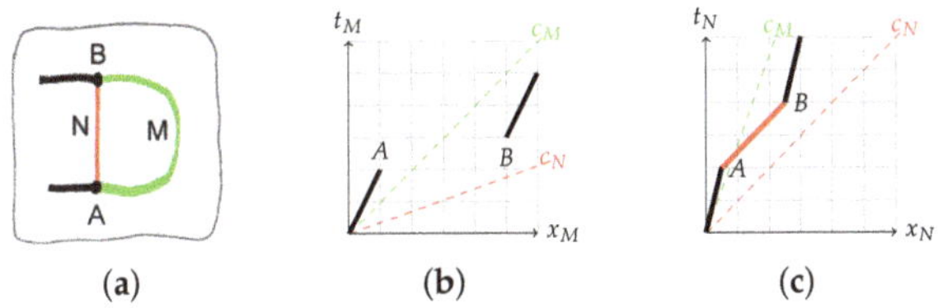

Figure 3. Schematic drawing of (**a**) worldlines of interdimensional "jump" motion, as seen from the outer, embedding space: (**a**) "dive" into N at A, reappearance at B; (**b**) space–time diagram as seen from intrinsic coordinates in M; (**c**) space–time diagram as seen from intrinsic coordinates in N.

The second example, depicted in Figure 4, shows an interdimensional "drag" motion that uses a dimensional motion in N whose velocity exceeds that of the normal signal velocity in M. As already mentioned, in both of these cases, consistency is guaranteed by the overall consistency in the outer embedding space.

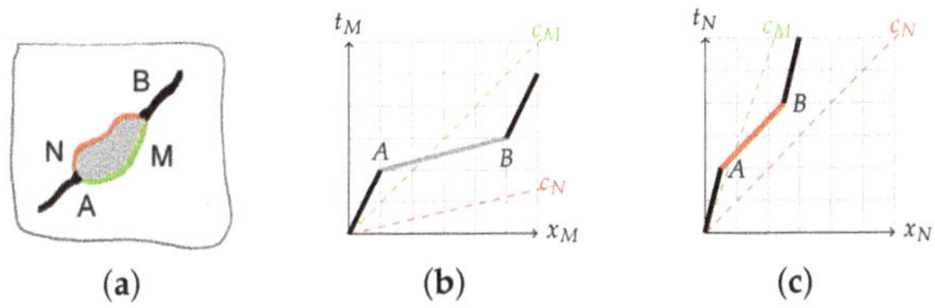

Figure 4. Schematic drawing of (**a**) worldlines of interdimensional forced, continuous motion, as seen from the outer, embedding space: (**a**) until A and from B, the motion is dominated by constraints on the velocity v_N, and between A and B, the velocity c_N dominates; (**b**) space–time diagram as seen from intrinsic coordinates in M; (**c**) space–time diagram as seen from intrinsic coordinates in N.

5. Further Speculations

Let us conclude this article with some speculative thoughts. The first is on limits to isolating the dimensions from one another, from "keeping them apart"; in particular, in the event of some catastrophic occurrence. It may well be that the domain of dimensional intersections may increase, as such events may dominate and spread to larger parts of the "outer" space.

Secondly, interdimensionality can be compared to computer simulations, with interfaces between such universes serving as intertwining regions. The difference between virtual reality (exchanges) and (intertwining) interdimensionality is the emphasis on measure-theoretic aspects in the latter case.

The matters discussed here must be considered highly speculative, and far from a fully developed formal theory. Nevertheless, it is our conviction that, to progress, science has to expand and explore a great variety of options, even if they appear remote to the contemporary mind.

Funding: This research was funded in whole, or in part, by the Austrian Science Fund (FWF), Project No. I 4579-N. For the purpose of open access, the author has applied a CC BY public copyright license to any Author Accepted Manuscript version arising from this submission.

Institutional Review Board Statement: Not applicable.

Informed Consent Statement: Not applicable.

Data Availability Statement: Not applicable.

Conflicts of Interest: The author declares no conflict of interest. The funders had no role in the design of the study; in the collection, analyses, or interpretation of data; in the writing of the manuscript, or in the decision to publish the results.

References

1. Gold, M.E. Language identification in the limit. *Inf. Control* **1967**, *10*, 447–474. [CrossRef]
2. Angluin, D.; Smith, C.H. Inductive Inference: Theory and Methods. *ACM Comput. Surv.* **1983**, *15*, 237–269. [CrossRef]
3. Frankfurt, H.G. *On Bullshit*; Princeton University Press: Princeton, NJ, USA, 2009.
4. Lakatos, I. Falsification and the methodology of scientific research programmes. In *The Methodology of Scientific Research Programmes. Philosophical Papers Volume 1*; Cambridge University Press: Cambridge, UK, 2012. [CrossRef]
5. Kuhn, T.S. *The Structure of Scientific Revolutions*, 4th ed.; University of Chicago Press: Chicago, IL, USA, 2012.
6. Feyerabend, P.K. *Problems of Empiricism. Philosophical Papers, Volume 2*; Cambridge University Press: Cambridge, UK, 1981. [CrossRef]
7. Feyerabend, P.K. How to Defend Society Against Science. *Radic. Philos.* **1975**, *11*, 3–9.
8. Svozil, K. Dimensional reduction via dimensional shadowing. *J. Phys. A Math. Gen.* **1986**, *19*, L1125–L1127. [CrossRef]
9. Falconer, K.J. *Fractal Geometry: Mathematical Foundations and Applications*, 3rd ed.; John Wiley & Sons: Chichester, UK, 2014.
10. Zeilinger, A.; Svozil, K. Measuring the dimension of space–time. *Phys. Rev. Lett.* **1985**, *54*, 2553–2555. [CrossRef] [PubMed]
11. Svozil, K.; Zeilinger, A. Dimension of space–time. *Int. J. Mod. Phys.* **1986**, *A1*, 971–990. [CrossRef]
12. Wertheimer, M. Experimental studies on seeing motion. In *On Perceived Motion and Figural Organization*; Spillmann, L., Ed.; The MIT Press: Cambridge, MA, USA, 2012; pp. 1–91.
13. Goldstein, E.B.; Brockmole, J.R. *Sensation and Perception*, 10th ed.; Cengage Learning: Boston, MA, USA, 2017.
14. Federer, H. *Geometric Measure Theory*; Springer: Berlin/Heidelberg, Germany, 1969. [CrossRef]
15. Wagon, S. The Banach-Tarski Paradox. In *Encyclopedia of Mathematics and Its Applications*; Cambridge University Press: Cambridge, UK, 1985. [CrossRef]
16. Abbott, E. *Flatland: A Romance of Many Dimensions*, unabridged 1884 ed.; Princeton University Press: Princeton, NJ, USA, 2015.
17. Svozil, K. Towards Fractal Gravity. *Found. Sci.* **2019**, *25*, 275–280. [CrossRef]
18. Hill, T.; Sanders, B.C.; Deng, H. Cooperative light scattering in any dimension. *Phys. Rev. A* **2017**, *95*, 033832. [CrossRef]
19. Mitic, V.V.; Lazovic, G.; Paunovic, V.; Cvetkovic, N.; Jovanovic, D.; Veljkovic, S.; Randjelovic, B.; Vlahovic, B. Fractal frontiers in microelectronic ceramic materials. *Ceram. Int.* **2019**, *45*, 9679–9685. [CrossRef]
20. Svozil, K. Relativizing Relativity. *Found. Phys.* **2000**, *30*, 1001–1016. [CrossRef]
21. Carr, B. (Ed.) *Universe or Multiverse?*; Cambridge University Press: Cambridge, UK, 2007. [CrossRef]
22. Lentz, E.W. Breaking the warp barrier: Hyper-fast solitons in Einstein-Maxwell-plasma theory. *Class. Quantum Grav.* **2021**, *38*, 075015. [CrossRef]

axioms MDPI

Article

On Turing Machines Deciding According to the Shortest Computations

Florin Manea [1,2]

[1] Computer Science Department, Göttingen University, 37073 Göttingen, Germany; florin.manea@cs.uni-goettingen.de
[2] Campus-Institut Data Science, Göttingen University, 37073 Göttingen, Germany
† Extended version of a paper of Proceedings of the Conference on Computability in Europe 2011.

Abstract: In this paper we propose and analyse from the computational complexity point of view several new variants of nondeterministic Turing machines. In the first such variant, a machine accepts a given input word if and only if one of its shortest possible computations on that word is accepting; on the other hand, the machine rejects the input word when all the shortest computations performed by the machine on that word are rejecting. We are able to show that the class of languages decided in polynomial time by such machines is $\mathbf{P}^{\mathbf{NP}[\log]}$. When we consider machines that decide a word according to the decision taken by the lexicographically first shortest computation, we obtain a new characterization of $\mathbf{P}^{\mathbf{NP}}$. A series of other ways of deciding a language with respect to the shortest computations of a Turing machine are also discussed.

Keywords: computational complexity; Turing machine; oracle Turing machine; shortest computations

Citation: Manea, F. On Turing Machines Deciding According to the Shortest Computations. *Axioms* **2021**, *10*, 304. https://doi.org/10.3390/axioms10040304

Academic Editor: Cristian S. Calude

Received: 13 October 2021
Accepted: 11 November 2021
Published: 13 November 2021

1. Introduction

The computation of a nondeterministic Turing machine and, in fact, any computation of a nondeterministic machine that consists of a sequence of moves can be represented as a (potentially infinite) tree. Each node of this tree is an instantaneous description (ID for short); that is, a string encoding the configuration of the machine at a given moment: the content of the machine's memory and the current state of the machine. The children of a node are the IDs encoding the possible configurations in which the machine can be found after a (nondeterministic) move is performed starting from the ID corresponding to that node. If the computation is finite then the tree is also finite and each leaf of the tree encodes a final ID: an ID in which the state is either accepting or rejecting. The machine accepts if and only if one of the leaves encodes the accepting state (also in the case of infinite trees), and rejects if the tree is finite and all the leaves encode the rejecting state.

Therefore, in the case of finite computations, one can check if a word is accepted/rejected by a machine by searching in the computation tree for a leaf that encodes an accepting ID. Theoretically, this is done by a simultaneous traversal of all the possible paths in the tree (as we can deduce, for instance, from the definition of the time complexity of a nondeterministic computation). However, in practice, it is done by traversing each path at a time, until an accepting ID is found, or until the whole tree was traversed. Unfortunately, this may be a very time consuming task. Consequently, one may be interested in heuristic methods that may speed up this search, or, in other words, methods of using nondeterministic machines in a more efficient manner.

Our paper proposes such a method: the machine accepts a word if and only if one of the shortest paths in the computation tree ends with an accepting ID and rejects the input word if all the shortest paths end with rejecting IDs. Intuitively, we traverse the computation tree on levels and, as soon as we reach a level containing a leaf, we check if there is a leaf encoding an accepting ID on that level, and accept, or if all the leaves on that level are rejecting IDs, and, consequently, reject. While it is not hard to see that the class of languages which are accepted (respectively decided) by these machines is the class of recursively enumerable languages (respectively, the class of recursive languages), we are able to show that the class of languages that are decided according to this strategy by Turing machines, whose shortest computations have a polynomial number of steps, equals the class $\mathbf{P}^{\mathbf{NP}[\log]}$. As a consequence of this result we can also show that the class of languages that are decided by Turing machines, working in nondeterministic polynomial time on any input but deciding according to the computations that have a minimal number of nondeterministic moves, also equals the class $\mathbf{P}^{\mathbf{NP}[\log]}$. These results continue a series of characterizations of $\mathbf{P}^{\mathbf{NP}[\log]}$, started in [1–3].

Then, we propose another method: the machine accepts (rejects) a word if and only if the the first leaf that we meet in a breadth-first-traversal of the computation tree encodes an accepting ID (respectively, encodes a rejecting ID); note that in this case, one must define first an order between the sons of a node in the computation tree. Again, it is not hard to show that these machines have the same computational power as unrestricted Turing machines. However, we show that, in the case of ordering the tree lexicographically, the class of languages that are decided, according to this new strategy, by Turing machines whose shortest computations have a polynomial number of steps, equals the class $\mathbf{P}^{\mathbf{NP}}$.

The research presented in this paper is related to a series of papers presenting variants of nondeterministic Turing machines, working in polynomial time, that accept (or reject) a word if and only if a specific property is (respectively, is not) verified by the possible computations of the machine on that word. We recall, for instance: polynomial machines that accept if and only if the number of accepting paths is even ($\oplus\mathbf{P}$ from [4]), polynomial machines which accept if at least $1/2$ of their computations are accepting, and reject if at least $1/2$ of their computations are rejecting (the class **PP** [5], which coincidentally include $\mathbf{P}^{\mathbf{NP}[\log]}$ [6]), or polynomial machines that accept if at least $2/3$ of the computation paths accept and reject if at most $1/3$ of the computation paths accept (the class of bounded-error probabilistic polynomial time $\mathbf{BPP}_{\mathbf{path}}$ from [7]); several other examples can be found on the Complexity Zoo web page (https://complexityzoo.net/ (accessed on 10 November 2021), a web page constructed and maintained, at the time when this paper was submitted, by the zookeeper Scott Aaronson, the veterinarian Greg Kuperberg, and the zoo conservationist Oliver Habryka on behalf of the LessWrong community) or in [8]. However, instead of looking at all the computations, we look just at the shortest ones, and instead of asking questions regarding the number of accepting/rejecting computations, we just ask existential questions about the shortest computations.

Our work finds motivations also in the area of nature-inspired supercomputing models. Some of these models (see [9,10], for instance) were shown to be complete by simulating, in a massively parallel manner, all the possible computations of a nondeterministic Turing machine; characterizations of several complexity classes, like **NP**, **P** and **PSPACE**, were obtained in this framework. However, these machines were, generally, used to accept languages, not to decide them; in the case when a deciding model was considered [9], the rejecting condition was just a mimic of the rejecting condition from classical computing models. Modifying such nature-inspired machines in order to decide as soon as a possible accepting/rejecting configuration is obtained, in one of the computations simulated in parallel, seems to be worth analysing: such a halting condition looks closer to what really happens in nature, and it leads to a reduced use of resources, comparing to the case when the machine kept on computing until all the possibilities were explored. Moreover, from a theoretical point of view, considering such halting conditions could lead to novel characterizations of a series of complexity classes (like the ones discussed in this paper) by means of nature-inspired computational models, as they seem quite close to the idea of deciding with respect to the shortest computations. To this end, we refer to the papers [11,12], and we leave open the question of whether similar results could be obtained for bio-inspired machines with more particular and compact structure [13–15] or for bio-inspired problem solvers [16].

2. Basic Definitions

The reader is referred to [8,17,18] for the basic definitions regarding Turing machines, oracle Turing machines, complexity classes and complete problems. In the following we present just the intuition behind these concepts, as a more detailed presentation would exceed the purpose of this paper.

A k-tape Turing machine is a construct $M = (Q, V, U, q_0, acc, rej, B, \delta)$, where Q is a finite set of states, q_0 is the initial state, acc and rej are the accepting state, respectively, the rejecting state, U is the working alphabet, B is the blank-symbol, V is the input alphabet and $\delta : (Q \setminus \{acc, rej\}) \times U^k \to 2^{(Q\times(U\setminus\{B\})^k\times\{L,R\}^k)}$ is the transition function (that defines the moves of the machine). An instantaneous description (ID for short) of a Turing machine is a word that encodes the state of the machine and the contents of the tapes (actually, the finite strings of nonblank symbols that exist on each tape), and the position of the tape heads, at a given moment of the computation. An ID is said to be final if the state encoded in it is the accepting or the rejecting state. A computation of a Turing machine on a given word can be described as a sequence of IDs: each ID is transformed into the next one by simulating a move of the machine. If the computation is finite then the associated sequence is also finite and it ends with a final ID; a computation is said to be an accepting (respectively, rejecting) one, if and only if the final ID encodes the accepting state (respectively, rejecting state). All the possible computations of a nondeterministic machine on a given word can be described as a (potentially infinite) tree of IDs: each ID is transformed into its sons by simulating the possible moves of the machine; this tree is called a computation tree.

A word is accepted by a Turing machine if there exists an accepting computation of the machine on that word; it is rejected if all the computations are rejecting. A language is accepted (decided) by a Turing machine if all its words are accepted by the Turing machine, and no other words are accepted by that machine (respectively, all the other words are rejected by that machine). The class of languages accepted by Turing machines is denoted by **RE** (and is known as the class of recursively enumerable languages), while the class of languages decided by Turing machines is denoted by **REC** (and called the class of recursive languages).

The time complexity (or length) of a finite computation on a given word is the minimum between the number of IDs that occur in an accepting computation of that word and the height of the computations-tree of the machine on the word. A language is said to be decided in polynomial time if there exists a Turing M machine and a polynomial f such that the time complexity of a computation of M on each word of length n is less than $f(n)$, and M accepts exactly the given language. The class of languages decided by deterministic Turing machines in polynomial time is denoted **P** and the class of languages decided by nondeterministic Turing machines in polynomial time is denoted **NP**. If a machine decides a language in polynomial time we usually say that this machine works in polynomial time.

A Turing machine with oracle A, where A is a language over the working alphabet of the machine, is a regular Turing machine that has a special tape (the oracle tape) and a special state (the query state). The oracle tape is just as any other tape of the machine, but, every time the machine enters the query state, a move of the machine consists of checking if the word found on the oracle tape is in A or not, and returning the answer.

We denote by $\mathbf{P}^{\mathbf{NP}}$ the class of languages decided by deterministic Turing machines, that work in polynomial time, with oracles from **NP**. We denote by $\mathbf{P}^{\mathbf{NP}[\log]}$ the class of languages decided by deterministic Turing machines, that work in polynomial time, with oracles from **NP**, and which can enter the query state at most $\mathcal{O}(\log n)$ times in a computation on a input word of length n.

The following problem is complete for $\mathbf{P}^{\mathbf{NP}}$, with respect to polynomial time reductions (see [19] for a proof):

Problem 1. *(Odd—Travelling Salesman Problem, TSP_{odd}) Let n be a natural number, and d be a function $d : \{1, \ldots, n\} \times \{1, \ldots, n\} \to \mathbb{N}$. Decide if the minimum value of the set $I = \{\sum_{i=1}^{n} d(\pi(i), \pi(i+1)) \mid \pi$ is a permutation of $\{1, \ldots, n\}$, and $\pi(n+1) = \pi(1)\}$ is odd.*

We assume that the input of this problem is given as the natural number n, and n^2 numbers representing the values $d(i,j)$, for all i and j. The size of the input is the number of bits needed to represent the values of d times n^2.

Next we describe a $\mathbf{P}^{\mathbf{NP}[\log]}$-complete problem; however, we need a few preliminary notions (see [20] for a detailed presentation). Let n be a natural number and let $C = \{c_1, \ldots, c_n\}$ be a set of n candidates. A preference order on C is an ordered list $\langle c_{\pi(1)} < c_{\pi(2)} < \ldots < c_{\pi(n)} \rangle$, where π is a permutation of $\{1, \ldots, n\}$; if c_i appears before c_j in the list we say that the candidate c_i is preferred to the candidate c_j in this order. Given a multiset V of preference orders on a set of n candidates C (usually V is given as a list of preference orders) we say that the candidate c_i is a Condorcet winner, with respect to the preference orders of V, if c_i is preferred to each other candidate in strictly more than half of the preference orders. We define the Dodgson score of a candidate c, with respect to V, as the smallest number of exchanges of two adjacent elements in the preference orders from V (switches, for short) needed to make c a Condorcet winner; we denote this score with $Score(C,c,V)$. In [20] it was shown that the following problem is $\mathbf{P}^{\mathbf{NP}[\log]}$-complete, with respect to polynomial time reductions:

Problem 2. *(Dodgson Ranking, DodRank) Let n be a natural number, let C be a set of n candidates, and c and d two candidates from C. Let V be a multiset of preference orders on C. Decide if $Score(C,c,V) \leq Score(C,d,V)$.*

We assume that the input of this problem is given as the natural number n, two numbers c and d less or equal to n, and a list of preference orders V, encoded as permutations of the set $\{1, \ldots, n\}$. If we denote by $\#(V)$ the number of preference orders in V, then the size of the input is $\mathcal{O}(\#(V) n \log n)$.

The connection between decision problems and languages is discussed in [18]. When we say that a decision problem is solved by a Turing machine, of certain type, we actually mean that the language corresponding to that decision problem is decided by that machine.

3. Shortest Computations

In this section we propose a modification of the way Turing machines decide an input word. Then we propose a series of results on the computational power of these machines and the computational complexity classes defined by them.

Definition 1. *Let M be a Turing machine and w be a word over the input alphabet of M. We say that w is accepted by M with respect to shortest computations if there exists at least one finite possible computation of M on w, and one of the shortest computations of M on w is accepting; w is rejected by M with regard to shortest computations if there exists at least one finite computation of M on w, and all the shortest computations of M on w are rejecting. We denote by $L_{sc}(M)$ the language accepted by M with regard to shortest computations, i.e., the set of all words accepted by M, with regard to shortest computations. We say that the language $L_{sc}(M)$ is decided by M with regard to shortest computations if all the words not accepted by M, with regard to shortest computations, are rejected with regard to shortest computations.*

The following remark shows that the computational power of the newly defined machines coincides with that of classic Turing machines.

Remark 1. *The class of languages accepted by Turing machines with regard to shortest computations equals* **RE**, *while the class of languages decided by Turing machines with regard to shortest computations equals* **REC**.

Proof. On the one hand, since any language from **REC** (respectively, **RE**) is decided (accepted) by a deterministic Turing machine, it is clear that it is also decided (accepted) with regard to shortest computations by the same machine. Indeed, a deterministic machine has a single computation, and this is also the shortest computation, so the decision reached on this computation is the same decision reached when the machine works according to the shortest computation policy.

On the other hand, if a language is decided (respectively, accepted) by a Turing machine M with regard to shortest computations then that language is decided (accepted) by a classic deterministic Turing machine M' as follows. The machine M' simply generates the computation tree of M on an input word w level by level. Basically, this is a very simple process: starting with all the configurations on one level, M' simulates one computational step of M on each of them, and collects the resulting configurations. In this way, M' explores, in order, the levels of the computation tree of M. Then, the machine M' stops as soon as it generates a level of the computation tree of M which contains a final ID. It accepts the input word if the respective level contains an accepting ID, and rejects otherwise. To a certain extent, the deterministic machine M' explores the computation tree of M breadth-first, and stops this exploration on the first level of this computation tree which contains final ID; the decision is then made by analysing the IDs of the respective level. □

Next we define a computational complexity measure for the Turing machines that decide the shortest computations.

Definition 2. *Let M be a Turing machine, and w be a word over the input alphabet of M. The time complexity of the computation of M on w, measured with regard to shortest computations, is the length of the shortest possible computation of M on w. A language L is said to be decided in polynomial time with regard to shortest computations if there exists a Turing M machine and a polynomial f such that the time complexity of a computation of M on each word of length n, measured with regard to shortest computations, is less than $f(n)$, and $L_{sc}(M) = L$. We denote by $PTime_{sc}$ the class of languages decided by Turing machines in polynomial time with regard to shortest computations.*

The main result of this section is the following:

Theorem 1. $PTime_{sc} = \mathbf{P}^{\mathbf{NP}[\log]}$.

Proof. The proof will be structured in two parts. First, we show the *upper bound* $PTime_{sc} \subseteq \mathbf{P}^{\mathbf{NP}[\log]}$, and then we show the *lower bound* $PTime_{sc} \supseteq \mathbf{P}^{\mathbf{NP}[\log]}$.

For the first part of the proof, let $L \subseteq V^*$ be a language in $PTime_{sc}$ and let M be a Turing machine that decides L in polynomial time with regard to shortest computations. Additionally, let f be a polynomial such that the time complexity of the computation of M on each word of length n, measured with regard to shortest computations, is less than $f(n)$. Finally, let # be a symbol not contained in V.

We define the language $L' = \{x\#w\#1^k \mid w \in V^*, x \in \{0,1\}$, and, if $x = 1$ (respectively, $x = 0$) there exists an accepting (respectively, rejecting) computation of M, of length less than k, on the input word $w\}$. It is not hard to see that L' is in **NP**. A nondeterministic machine deciding L' works as follows: it simulates, nondeterministically, a computation of at most k steps of M, and accepts if and only if $x = 1$, or, respectively, $x = 0$, and the simulated computation is accepting, or, respectively, rejecting; otherwise (i.e., if in the k steps simulated by the machine a final configuration was not obtained) it rejects. Clearly, this machine works in polynomial time.

A deterministic Turing machine M', with oracle L', accepting L implements the following strategy, on an input word w:

1. M' searches (by binary search) the minimum length of an accepting computation of M on w, with length less or equal to $f(|w|)$. In this search, the machine queries the oracle L' for $\mathcal{O}(\log_2(f(|w|)))$ times, asking, in each of these queries, if a string of the form $1\#w\#1^k$, with $k \le f(|w|)$, is in L'.
2. Let n_0 be the minimum length of an accepting computation, with length less than or equal to $f(|w|)$, computed in the previous step (we assume that n_0 is set to a special value, $f(n) + 1$ for instance, if the search is unsuccessful). The machine verifies now,

by another oracle query, if $0\#w\#1^{n_0-1} \in L'$ (i.e., if there exists a shorter rejecting computation of M). If the answer of the last query is positive, M' rejects the input word, otherwise, it accepts.

Since the machine M has at least one possible computation on w of length less than $f(|w|)$, and that $w \in L$ if and only if the shortest computation of M accepts, it is clear that the machine M' decides the language L. Furthermore, M' works in polynomial time and makes at most $\mathcal{O}(\log n)$ queries to the oracle L'; therefore, $L \in \mathbf{P}^{\mathbf{NP}[\log]}$. This completes the proof of the upper bound.

For the second inclusion, note that the class $PTime_{sc}$ is closed to polynomial-time reductions. That is, if $L \in PTime_{sc}$ and L' is polynomial-time reducible to L, then $L' \in PTime_{sc}$. Indeed, assume that g is a function, that can be computed in polynomial time by a deterministic Turing machine such that, $w \in L'$ if and only if $g(w) \in L$. A machine that decides with regard to shortest computations the language L' works as follows: first, for the input w, it computes deterministically the function $g(w)$, and, then, runs the machine accepting L on the input $g(w)$; it is clear that this machine implements the desired behaviour, and that it works in polynomial time, measured with regard to shortest computations. Therefore, it is sufficient to show that the $\mathbf{P}^{\mathbf{NP}[\log]}$-complete problem *DodRan* can be solved in polynomial time by a Turing machine M that makes a decision with regard to shortest computations.

Let us first make several denotations. The input of M consists in the number n, the set C of n candidates, c and d two candidates from C, and V the multiset of preference orders on C (encoded as explained in the previous section). It is not hard to see that one can verify if a candidate is a Condorcet winner for the multiset V of preference orders on C in polynomial time; let f be a polynomial that upper bounds the time needed to do this checking, for every n and $\#(V)$. Note that one needs at most $(n-1)\left(\left\lfloor\frac{\#(V)}{2}\right\rfloor+1\right)$ switches to make a candidate a Condorcet winner, since, in the worst case, we must bring this candidate from the last position to the first position in $\left\lfloor\frac{\#(V)}{2}\right\rfloor+1$ of the orders. Additionally, making $(n-1)\left(\left\lfloor\frac{\#(V)}{2}\right\rfloor+1\right)$ switches in the orders of V requires polynomial time. Let g be a polynomial that sets the upper bounds for the time needed to make $(n-1)\left(\left\lfloor\frac{\#(V)}{2}\right\rfloor+1\right)$ switches, for every n and $\#(V)$.

This machine implements the following algorithm:

1. M writes, nondeterministically, two numbers k_1 and k_2 (as the strings 1^{k_1} and 1^{k_2}), with $k_i \leq (n-1)\left(\left\lfloor\frac{\#(V)}{2}\right\rfloor+1\right)$ for $i \in \{1,2\}$. Then, M chooses nondeterministically k_1 switches to be made in V, and saves them as the set T_1, and k_2 switches to be made in V, and saves them as the set T_2.
2. M makes (deterministically) the switches from T_1, and saves the newly obtained preference orders as a multiset V_1. M makes (deterministically) the switches from T_2, and saves the newly obtained preference orders as a multiset V_2.
3. M checks (deterministically) if c is a Condorcet winner in V_1. If the answer is positive it goes to step 4, otherwise it makes $2f(n,\#(V)) + 2g(n,\#(V))$ dummy steps and rejects the input word.
4. M checks (deterministically) if d is a Condorcet winner in V_2. If the answer is positive it goes to step 7, otherwise it makes $2f(n,\#(V)) + 2g(n,\#(V))$ dummy steps and rejects the input word.
5. If $k_1 \leq k_2$ the machine accepts the input, otherwise it rejects it.

First, let us see that M works correctly. In step 1 it chooses nondeterministically some switches in V, that are supposed to make c and d Condorcet winners, respectively. Notice that the length of a possible computation performed in this step depends on the choice of the numbers k_1 and k_2; if these numbers are smaller, then the computation is shorter. Then in step 2 the machine actually makes (deterministically) the switches chosen in the previous step. The length of a possible computation, until this moment, is still determined by the choice of k_1 and k_2. In steps 3 and 4 the machine verifies if those switches were indeed good to make c and d winners, according to the orders modified by the previously chosen moves. If they were both transformed in winners by the chosen switches, the computation continues with to step 5; otherwise, the machine makes a sequence of dummy steps, long enough to make that computation irrelevant for the final answer of the machine on the given input. Note that at least one choice of the switches, in step 1, makes both c and d winners. Now, the shortest computations are those ones in which both c and d were transformed into winners and the chosen numbers k_1 and k_2 are minimal. Yet this is exactly the case when $k_1 = Score(C, c, V)$ and $k_2 = Score(C, d, V)$. In the step 5, all the computations in which c and d were transformed into winners are completed by a deterministic comparison between k_1 and k_2. Thus, after the execution of this step the shortest computations remain the ones where $k_1 = Score(C, c, V)$ and $k_2 = Score(C, d, V)$; the decision of this computation is to accept, if $k_1 \leq k_2$, or to reject, otherwise. Consequently, M accepts if and only if $Score(C, c, V) \leq Score(C, d, V)$, and rejects otherwise. Moreover, it is rather easy to see that M works in polynomial time, since each of the 5 steps described above can be completed in polynomial time.

In conclusion, we showed that *DodRan* can be solved in polynomial time by a Turing machine that decides with regard to shortest computations. It follows that $PTime_{sc} \supseteq \mathbf{P}^{\mathbf{NP}[\log]}$, and this ends our proof. □

The technique used in the previous proof to show that $\mathbf{P}^{\mathbf{NP}[\log]}$-complete problems can be solved in polynomial time by Turing machines that decide with regard to shortest computations suggests another characterization of $\mathbf{P}^{\mathbf{NP}[\log]}$. In this respect, consider nondeterministic Turing machines, working in polynomial time, that decide an input according to the decisions of the computations in which the least number of nondeterministic moves is made. Such a machine can be formally defined as follows:

Definition 3. *Let M be a Turing machine working in polynomial time and w be a word over the input alphabet of M. We say that w is accepted by M with respect to the computations with a minimum number of nondeterministic moves if one of the possible computations of M on w, in which M makes the minimum number of nondeterministic moves, is accepting; w is rejected by M with regard to the computations with minimum number of nondeterministic moves if all the possible computations of M on w, in which M makes the minimum number of nondeterministic moves, are rejecting.We denote by $L_{nm}(M)$ the language decided by M with regard to the computations with a minimum number of nondeterministic moves and by $PTime_{nm}$ the class of all the languages decided in this manner.*

It is not hard to see that, given a Turing machine working in polynomial time and an input word for that machine, the machine will always decide the input word with regard to the computations with minimum number of nondeterministic moves, since all of its computations are finite. One can show the following result.

Theorem 2. $PTime_{nm} = \mathbf{P}^{\mathbf{NP}[\log]}$.

Proof. We can use a proof similar to the one of Theorem 1.

For the inclusion $PTime_{nm} \subseteq \mathbf{P}^{\mathbf{NP}[\log]}$ we can assume, without loss of generality, that the machine accepting a language from $PTime_{nm}$ has all the possible computations on an input of length n of the same length $f(n)$, for some polynomial f (we can complete some of the computations with dummy deterministic steps, in order to make this happen). Then we just have to search (using binary search) for the computation with the minimum number of nondeterministic moves, and check if it is an accepting or rejecting one.

For the inclusion $PTime_{nm} \supseteq \mathbf{P}^{\mathbf{NP}[\log]}$, we use the machine constructed in the proof of $PTime_{sc} \supseteq \mathbf{P}^{\mathbf{NP}[\log]}$, and note that the shortest computations performed by this machine on a certain input are also the computations where the minimum number of nondeterministic moves are made. This concludes our proof. □

4. The First Shortest Computation

4.1. Ordered Turing Machines

In the previous section we proposed a decision mechanism of Turing machines that basically consisted in identifying the shortest computations of a machine on an input word, and checking if one of these computations is an accepting one, or not. Now we analyse how the properties of the model are changed if we order the computations of a machine and the decision is made according to the first shortest computation, in the defined order.

Let $M = (Q, V, U, q_0, acc, rej, B, \delta)$ be a t-tape Turing machine, and assume that $\delta(q, a_1, \ldots, a_t)$ is a totally ordered set, for all $a_i \in U$, $i \in \{1, \ldots, t\}$, and $q \in Q$; we call such a machine an *ordered Turing machine*. Let w be a word over the input alphabet of M. Assume that s_1 and s_2 are two (potentially infinite) sequences describing two possible computations of M on w. We say that s_1 is lexicographically smaller than s_2 if s_1 has fewer moves than s_2, or they have the same number of steps (potentially infinite), the first k IDs of the two computations coincide and the transition that transforms the kth ID of s_1 into the $k+1$th ID of s_1 is smaller than the transition that transforms the kth ID of s_2 into the $k+1$th ID of s_2, with respect to the predefined order of the transitions. It is not hard to see that this is a total order on the computations of M on w. Therefore, given a finite set of computations of M on w, one can define the lexicographically first computation of the set as that one which is lexicographically smaller than all the others.

Definition 4. *Let M be an ordered Turing machine, and w be a word over the input alphabet of M. We say that w is accepted by M with respect to the lexicographically first computation if there exists at least one finite possible computation of M on w, and the lexicographically first computation of M on w is accepting; w is rejected by M with regard to the lexicographically first computation if the lexicographically first computation of M on w is rejecting. We denote by $L_{lex}(M)$ the language accepted by M with regard to the lexicographically first computation. We say that the language $L_{lex}(M)$ is decided by M with regard to the lexicographically first computation if all the words not contained in $L_{lex}(M)$ are rejected by M.*

As in the case of Turing machines that decide with regard to shortest computations, the class of languages accepted by Turing machines with regard to the lexicographically first computation equals **RE**, while the class of languages decided by Turing machines with regard to the lexicographically first computation equals **REC**. The time complexity of the computations of Turing machines that decide with regard to the lexicographically first computation is defined exactly as in the case of machines that decide with regard to shortest computations. We denote by $PTime_{lex}$ *the class of languages decided by Turing machines in polynomial time with regard to the lexicographically first computation*. In this context, we are able to show the following theorem.

Theorem 3. $PTime_{lex} = \mathbf{P}^{\mathbf{NP}}$.

Proof. In the first part of the proof we show that $PTime_{lex} \subseteq \mathbf{P}^{\mathbf{NP}}$. Let L be a language in $PTime_{lex}$ and let M be a Turing machine that decides L in polynomial time with regard to the lexicographically first computation. Additionally, let f be a polynomial such that the time complexity of the computation of M on each word of length n, measured with regard to the lexicographically first computation, is less than $f(n)$.

We define the language $L' = \{x\#w\#w'\#1^k \mid w \in V^*, w'$ is a sequence of consecutive IDs of M, $x \in \{0,1\}$, and, if $x = 1$ (respectively, $x = 0$) there exists an accepting (respectively, rejecting) computation of M on the input word w of length less than k, starting with the sequence of IDs $w'\}$. It is not hard to see that L' is in **NP**. A nondeterministic machine deciding it works as follows: it simulates, nondeterministically, a computation of at most k steps of M, starting with the IDs in the sequence w', and accepts if and only if this $x = 1$,or, respectively, $x = 0$, and the simulated computation is accepting, or, respectively, rejecting; otherwise (i.e., if in the simulated computation steps a final configuration was not obtained), it rejects. Clearly, this machine works in polynomial time.

A deterministic Turing machine M', with oracle L', accepting L implements the following strategy, on an input word w:

1. M' searches (by binary search) the minimum length of a computation of M on w, with length less than or equal to $f(|w|)$. In this search, the machine queries the oracle L' for $\mathcal{O}(\log_2(f(|w|)))$ times, asking, in each of these queries, if a string of the form $1\#w\#\epsilon\#1^k$ and $0\#w\#\epsilon\#1^k$, with $k \leq f(|w|)$, is in L'. Let n_0 be the minimum length of a computation, with length less than or equal to $f(|w|)$.
2. Next, M' tries to construct, ID by ID, the first (shortest) computation of length n_0, using the oracle L'. Assume that w' is a sequence of IDs identified until a given moment as a prefix of the sequence encoding the first computation of length n_0, and we try to lengthen this sequence. Assume that $w_1, w_2, \ldots, w_k$ are the IDs that can be obtained from the last ID of w', ordered according to the transitions that were used to obtain them. We search the minimum i, with $1 \leq 1 \leq k$, such that $0\#w\#w'w_i\#1^{n_0}$ or $1\#w\#w'w_i\#1^{n_0}$ is in L'. Once we have identified this minimum value, denoted i_0, we add the ID w_{i_0} to the sequence w', and repeat the process described above, until w' contains n_0 IDs.
3. The machine finally checks if the string $1\#w\#w'\#1^{n_0}$ is in L', and if it is so accepts, or, if the string $0\#w\#w'\#1^{n_0}$ is in L', and, in this case, rejects.

It is not hard to see that M' correctly computes the length n_0 of the shortest computation of M on an input word w. Also, once this length computed, the first shortest computation is identified, and the machine checks if this computation is an accepting or a rejecting one. Thus, M' implements the desired behaviour. Finally, note that M' works in polynomial time: in step 2 it makes $\mathcal{O}(n_0)$ queries, asking if strings of polynomial length are in L', while the rest of the computation is clearly carried out in polynomial time. This completes the proof of the upper bound on $PTime_{lex}$.

To show the second inclusion, note that, similar to the case of machines deciding with regard to shortest computations, the class $PTime_{lex}$ is closed to polynomial-time reductions. Thus, it is sufficient to show that the $\mathbf{P}^{\mathbf{NP}}$-complete problem TSP_{odd} can be solved in polynomial time by a Turing machine M that decides with regard to the lexicographically first computation.

Therefore, we construct a Turing machine M that solves TSP_{odd} with regard to the lexicographically first computation. The input of this machine consists in a natural number n, and n^2 natural numbers, encoding the values of the function $d : \{1, \ldots, n\} \times \{1, \ldots, n\} \to N$. We can assume, without losing generality, that all the input numbers are given as decimal numbers; furthermore, we assume that all the n^2 numbers, that encode the values of the function d, have the same number of decimal digits, denoted by m (we may add some leading zeros at the beginning of these numbers in order to make this assumption hold). Therefore, the size of the input is $\mathcal{O}(n^2m)$. Additionally, let us make the assumption that every time we sum up n numbers of m digits we make exactly $f(m, n)$ steps, where f is a polynomial, and the sum is always represented using the same number of digits (clearly bounded by the input size).

This machine implements the following algorithm:

1. M writes, nondeterministically, a permutation π of $\{1, \ldots, n\}$ and computes, deterministically, the sum $S = \sum_{i=1}^{n} d(\pi(i), \pi(i+1))$. Let k be the number of digits of S.
2. M writes, nondeterministically, a number S_0 of k digits; this number may have some leading zeros. We assume that this step is performed in k computational steps, each consisting in choosing one of the moves $\{m_0, m_1, \ldots, m_9\}$ in which one of the digits $0, \ldots, 9$, respectively, is written. These moves are ordered $m_0 < m_1 < \ldots < m_8 < m_9$.
3. M writes, nondeterministically, a permutation π' of $\{1, \ldots, n\}$ and computes, deterministically, the sum $S' = \sum_{i=1}^{n} d(\pi'(i), \pi'(i+1))$.

4. M checks, deterministically, if $S' = S_0$. If yes, it goes to step 5, otherwise it makes $2n^2m$ dummy step and rejects.
5. M checks, deterministically, if S' is odd. If yes, it accepts, otherwise it rejects.

It is important to state that the order of the nondeterministic moves that are executed in steps 1 and 3 has no impact on the computation. For uniformity we consider that they are ordered, but we do not make any assumption on what order is actually used.

Before showing that the machine works correctly, we notice that it works in polynomial time. Indeed, it is not hard to see that every possible computation of M consists of a sequence of steps of polynomial length, and always ends with a decision.

To show the soundness of our construction, let us observe that all the possible computations implemented by the first 3 steps of the above algorithm have the same length. In the first of these steps we choose a possible permutation π of $\{1, \ldots, n\}$ and compute the sum $S = \sum_{i=1}^{n} d(\pi(i), \pi(i+1))$; in this way we have computed a possible solution of the Travelling Salesman Problem, defined by the function d, and the real solution of the problem should be at most S. Then we try to find another permutation π' that leads to a smaller sum. For this we choose first a number S_0 that has as many digits as S (of course, it may have several leading zeros); however, the computations are ordered in such a manner that a computation in which smaller numbers are constructed comes before a computation in which a greater number is constructed. Then, in steps 3 and 4, M verifies if S' can be equal to the sum $\sum_{i=1}^{n} d(\pi'(i), \pi'(i+1))$, for a permutation π' nondeterministically chosen. If the answer is yes then it means that S_0 is also a possible solution of the problem; otherwise, we conclude that the nondeterministic choices made so far were not really the good ones, so we reject after we make a long enough sequence of dummy steps, in order not to influence the decision of the machine. Finally, we verify if S_0 is odd, and accept if and only if this condition holds. By the considerations made above, it is clear that in all the shortest computations we identified some numbers that can represent solutions of the Travelling Salesman Problem; moreover, in the first of the shortest computations we have identified the smallest such number, i.e., the real solution of the problem. Consequently, the decision of the machine is to accept or to reject the input according to the parity of the solution identified in the first shortest computation, which is correct.

Summarizing, we showed that TSP_{odd} can be solved in polynomial time by a Turing machine that decides with regard to the lexicographically first computation. It follows that $PTime_{lex} \supseteq \mathbf{P}^{\mathbf{NP}}$, and this concludes our proof. □

Remark 2. *Note that the proof of Theorem 1 shows that $\mathbf{P}^{\mathbf{NP}[\log]}$ can be also characterized as the class of languages that can be decided in polynomial time with regard to shortest computations by nondeterministic Turing machines whose shortest computations are either all accepting or all rejecting. On the other hand, in the proof of Theorem 3, the machine that we construct to solve with regard to the lexicographically first computation the TSP_{odd} problem may have both accepting and rejecting shortest computations on the same input. This shows that $\mathbf{P}^{\mathbf{NP}[\log]} = \mathbf{P}^{\mathbf{NP}}$ if and only if all the languages in $\mathbf{P}^{\mathbf{NP}}$ can be decided with regard to shortest computations by nondeterministic Turing machines whose shortest computations on a given input are either all accepting or all rejecting.*

4.2. Ordering Functions

There is a point where the definition of the ordered Turing machine does not seem satisfactory: each time a machine has to execute a nondeterministic move, for a certain state and a tuple of scanned symbols, the order of the possible moves is the same, regardless of the input word and the computation performed until that moment. Therefore, we consider another variant of ordered Turing machines, in which such information is considered:
Let M be a Turing machine. We denote by $\langle M \rangle$ a binary encoding of this machine (see, for instance, [18]). It is clear that the length of the string $\langle M \rangle$ is a polynomial with respect to the number of states and the working alphabet of the machine M. Let $g : \{0,1,\#\}^* \to \{0,1,\#\}^*$ be a function such that $g(\langle M \rangle \# w_1 \# w_2 \# \ldots \# w_k) = w'_1 \# w'_2 \# \ldots \# w'_p$, given that $w_1, \ldots, w_k$ are

binary encodings of the IDs that appear in a computation of length k of M (we assume that they appear in this order, and that w_1 is an initial configuration), and $w'_1, \ldots, w'_p$ are the IDs that can be obtained in one move from w_k. Clearly, this function induces canonically an ordering on the computations of a Turing machine. Assume s_1 and s_2 are two (potentially infinite) sequences describing two possible computations of M on w. We say that s_1 is g-smaller than s_2 if the first k IDs of the two computations, which can be encoded by the strings $w_1, \ldots, w_k$, coincide, and $g(\langle M\rangle \# w_1 \# w_2 \# \ldots \# w_k) = w'_1 \# w'_2 \# \ldots \# w'_p$, the $k+1$th ID of s_1 is encoded by w'_i, the $k+1$th ID of s_2 is encoded by w'_j, and $i < j$. It is not hard to see that g induces a total order on the computations of M on w; thus, we will call such a function *an ordering function*. Therefore, given a finite set of computations of M on w we can define the g-first computation of the set as the one that is g-smaller than all the others.

Definition 5. *Let M be a Turing machine, and $g : \{0,1,\#\}^* \to \{0,1,\#\}^*$ be an ordering function. We say that w is accepted by M with respect to the g-first shortest computation if there exists at least one finite possible computation of M on w, and the g-first of the shortest computations of M on w is an accepting one; w is rejected by M with regard to the lexicographically first computation if the g-first shortest computation of M on w is a rejecting computation. We denote by $L^g_{fsc}(M)$ the language accepted by M with regard to the g-first shortest computation, i.e., the set of all words accepted by M, with regard to the g-first shortest computation. As in the case of regular Turing machines, we say that the language $L^g_{fsc}(M)$ is decided by M with regard to the g-first shortest computation if all the words not contained in $L^g_{fsc}(M)$ are rejected by that machine, with regard to the g-first shortest computation.*

It is not surprising that, if g is Turing computable, the class of languages accepted by Turing machines with regard to the g-first shortest computation equals **RE**, while the class of languages decided by Turing machines with regard to the lexicographically first computation equals **REC**. The time complexity of the computations of Turing machines that decide with regard to the g-first shortest computation is defined exactly as in the case of machines that decide with regard to shortest computations. We denote by $PTime^g_{fsc}$ *the class of languages decided by Turing machines in polynomial time with regard to the g-first shortest computation*. We also denote by $PTime_{ofsc}$ the union of all the classes $PTime^g_{fsc}$, where the ordering function g can be computed in polynomial deterministic time. We are now able to show the following theorem.

Theorem 4. $PTime_{ofsc} = \mathbf{P}^{\mathbf{NP}}$.

Proof. In fact, we will show that $PTime_{ofsc} = PTime_{lex}$. First, let us observe that the inclusion $PTime_{ofsc} \supseteq PTime_{lex}$ holds canonically. Indeed, the lexicographical order of the computations defined in the previous section is just a particular case of an order defined by an ordering function computable in deterministic polynomial time.

Further, we show that $PTime_{ofsc} \subseteq PTime_{lex}$. Given g an ordering function that can be computed in deterministic polynomial time, let L be a language and M be a Turing machine that decides in polynomial time L with regard to the g-first shortest computation. Let us assume, without loss of generality, that the time needed to compute the value of g for a string of k configurations of M, all having the same initial configuration, regardless of the configurations. We define an ordered machine M' and show that it decides L with regard to the lexicographically first computation, also in polynomial time.

We will not give the details of the construction of M', as they can be quite tedious, but we will give the main idea implemented by this machine. The machine M' basically simulates the computation of the machine M and keeps on a track (called "memory track") the encoding of M and the encodings of IDs of M that were obtained during the simulated computation. Assume that M' should simulate a move of M, provided that the current state of M is q and the scanned symbols are $(a_1, \ldots, a_k)$. First, M' enters in a state q_g in which it computes the value of the function g having as argument the string saved on the memory track. Suppose that the computed value is the string $w'_1\#w'_2\#\ldots\#w'_p$, and the machine M must make the transition m_i to obtain the ID w'_i from the current ID, for $i \in \{1,\ldots,p\}$. Accordingly, the machine M' enters in a state $q_{m_1,\ldots,m_p}$, and from this state it must make a nondeterministic move that simulates the move of M. However, we define M' such that its possibilities, in this case, are ordered: the first comes the move m_1, then the move m_2, and so on, finally coming m_p ($m_1 < m_2 < \ldots < m_p$, in the formalism of ordered machines). Once the move is simulated, the machine M' saves the encoding of the current ID of the simulated machine (again, we may assume that this operation can be done in the same time for any ID, since their length is bounded by a polynomial), and goes on to simulate the next move of M.

It is not hard to see that M' simulates soundly the behaviour of M. Basically, M' keeps a history of the computation performed by M and uses a subroutine, computing the function g, to ensure that the lexicographical order of the simulated computations coincides with the order defined by the function g for the machine M and its real computations. Additionally, the part of the algorithm implemented by M' that is not involved in the actual simulation (that is in keeping the history of the simulated computation and computing the values of g) depends only on the number of steps of M simulated until that point and on the input word, so it is quite easy to see that the shortest computations of M are simulated by the shortest computations of M'; moreover, the g-first shortest computation of M is simulated by the lexicographically first shortest computation of M'.

It follows that the language L is decided by M' in polynomial time with regard to the lexicographically first computation.

To conclude, we showed that $PTime_{ofsc} \subseteq PTime_{lex}$.

It follows that $PTime_{ofsc} = PTime_{lex}$ and, according to Theorem 3, we obtain the identity $PTime_{ofsc} = \mathbf{P}^{\mathbf{NP}}$. □

Notice that $\mathbf{P}^{\mathbf{NP}[\log]} \subseteq PTime^{g}_{fsc} \subseteq \mathbf{P}^{\mathbf{NP}}$, for all the ordering functions g which can be computed in polynomial deterministic time. The second inclusion is immediate from the previous Theorem, while the first one follows from the fact that any language in $\mathbf{P}^{\mathbf{NP}[\log]}$ is accepted with regard to shortest computations, in polynomial time, by a nondeterministic Turing machine whose shortest computations are either all accepting or all rejecting; clearly, the same machine can be used to show that the given language is in $PTime^{g}_{fsc}$.

It is interesting to see that for some particular ordering functions, as for instance the one that defines the lexicographical order discussed previously, a stronger result holds: $PTime^{g}_{fsc} = \mathbf{P}^{\mathbf{NP}}$ (where g is the ordering function). We leave as an open problem to see if this relation holds for all the ordering functions, or, if not, to see when it hold.

5. Conclusions and Further Work

In this paper, we have shown that considering a variant of Turing machine, that decides an input word according to the decisions of the shortest computations of the machine on that word, leads to new characterizations of two well-studied complexity classes $\mathbf{P}^{\mathbf{NP}[\log]}$ and $\mathbf{P}^{\mathbf{NP}}$. These results seem interesting since they provide alternative definitions of these two classes that do not make use of any other notion than the Turing machine (such as oracles, reductions, etc.) Note that some of our proofs rely on showing that complete problems can be solved by machines deciding with respect to the shortest computations. These complete problems were chosen according to the personal preferences of the author; clearly, other complete problems could have been solved similarly with the respective techniques. We feel, however, that our solutions capture entirely the ideas that connect the different complexity classes we characterize with the usage of shortest computations.

From a theoretical point of view, an attractive continuation of the present work would be to analyse if the equality results in Theorems 1–3 relativise. It is not hard to see that the upper bounds shown in these proofs are true even if we allow all the machines to have access to an arbitrary oracle. It remains to be settled if a similar result holds in the case of the lower bounds. However, we conjecture that the lower bounds do not hold in the presence of arbitrary oracles, highlighting, in this way, the difference between the way our variant of Turing machine decides and the way regular oracle Turing machines decide.

Nevertheless, other accepting/rejecting conditions related to the shortest computations could be investigated. As we mentioned in the Introduction, several variants of Turing machines that decide a word according to the number of accepting, or rejecting, computations were already studied. We intend to analyse what happens if we use similar conditions for the shortest computations of a Turing machine. In this respect, using the ideas of the proof of Theorem 3, one can show that

Theorem 5. *Given a nondeterministic polynomial Turing machine M_1, one can construct a nondeterministic polynomial Turing machine, with access to* **NP***-oracle, M_2, whose computations on an input word correspond bijectively to the short computations of M_1 on the same word, such that two corresponding computations are both either accepting, or rejecting.*

Proof. Let M be a nondeterministic Turing machine working in polynomial time. Furthermore, let f be a polynomial such that the time complexity of the computation of M on each word of length n is less than $f(n)$.

Recall the language $L' = \{x\#w\#w'\#1^k \mid w \in V^*, w'$ is a sequence of consecutive IDs of M, $x \in \{0,1\}$, and, if $x = 1$ (respectively, $x = 0$) there exists an accepting (respectively, rejecting) computation of M on the input word w of length less than k, starting with the sequence of IDs $w'\}$, from the proof of Theorem 3. Additionally, recall that L' is in **NP**.

We construct now a nondeterministic Turing machine M', with oracle L', that acts as follows:

1. M' searches (by binary search) the minimum length of a computation of M on w, with length less or equal to $f(|w|)$. In this search, the machine queries the oracle L' for $\mathcal{O}(\log_2(f(|w|)))$ times, asking, in each of these queries, if a string of the form $1\#w\#\epsilon\#1^k$ and $0\#w\#\epsilon\#1^k$, with $k \le f(|w|)$, is in L'. Let n_0 be the minimum length of a computation, with length less or equal to $f(|w|)$. This step is executed deterministically.
2. Next M' tries to construct nondeterministically, ID by ID, one of the shortest computations of M on w (the length of this computation is n_0), using the oracle L'. Assume that w' is a sequence of IDs identified until a given moment as a prefix of the sequence encoding such a computation, and we try to lengthen this sequence. Assume that $w_1, w_2, \dots, w_k$ are the IDs that can be obtained from the last ID of w'. We search all the possible i, with $1 \le 1 \le k$, such that $0\#w\#w'w_i\#1^{n_0}$ or $1\#w\#w'w_i\#1^{n_0}$ is in L'. Once we have identified these values, denoted by $i_1, \dots, i_p$, we add, nondeterministically, one of the IDs w_{i_j}, with $j \in \{1, \dots, p\}$ to the sequence w', and repeat the process described above, until w' contains n_0 IDs.
3. The machine finally checks if the string $1\#w\#w'\#1^{n_0}$, for a w' obtained in one of the possible computations, is in L', and if it is so, the computation is accepting, or, if the string $0\#w\#w'\#1^{n_0}$, for a w' obtained in one of the possible computations, is in L', and, in this case, the computation is rejecting.

It is not hard to see that M' correctly computes the length n_0 of the shortest computation of M on an input word w. Additionally, once this length computed, the shortest computations of M' are identified, and the machine simulates these computations nondeterministically. Thus, the computations of M can be put in a bijective correspondence with the shortest computations of M': one of the shortest computations of M corresponds to the computation of M' that simulates this shortest computation. Finally, note that M' works in nondeterministic polynomial time.

This concludes the proof of Theorem 5. □

This Theorem is useful to show upper bounds on the complexity classes defined by counting the accepting/rejecting shortest computations. Some examples in this direction are: $\mathbf{PP_{sc}} \subseteq \mathbf{PP^{NP}}$ (where $\mathbf{PP_{sc}}$ is the class of decision problems solvable by a nondeterministic polynomial Turing machine which accepts if and only if at least $1/2$ of the shortest computations are accepting, and rejects otherwise) or $\mathbf{BPP_{sc}} \subseteq \mathbf{BPP^{NP}_{path}}$ (where $\mathbf{BPP_{sc}}$ is the class of decision problems solvable by a nondeterministic polynomial Turing machine which accepts if at least $2/3$ of the shortest computations are accepting, and rejects if at least $2/3$ of the shortest computations are rejecting).

Remark 3. *However, in some cases, one can show stronger upper bounds; for instance,* $\mathbf{PP_{sc}} \subseteq \mathbf{PP^{NP[\log]}_{ctree}}$ *(where* $\mathbf{PP^{NP[\log]}_{ctree}}$ *is the class of decision problems solvable by a* **PP***-machine which can make a total number of* $\mathcal{O}(\log n)$ *queries to an* **NP***-language in its entire computation tree, on an input of length* n*). It seems an interesting problem to find lower bounds for such classes, as well.*

Proof. Let M be a nondeterministic Turing machine working in polynomial time. Additionally, let f be a polynomial such that the time complexity of the computation of M on each word of length n is less than $f(n)$.

Recall the language $L' = \{x\#w\#w'\#1^k \mid w \in V^*, w'$ is a sequence of consecutive IDs of M, $x \in \{0,1\}$, and, if $x = 1$ (respectively, $x = 0$), there exists an accepting (respectively, rejecting) computation of M on the input word w of length less than k, starting with the sequence of IDs $w'\}$, from the proof of Theorem 3. Recall also that L' is in **NP**.

We construct now a nondeterministic Turing machine M', with oracle L', that acts as follows:

1. M' searches (by binary search) the minimum length of a computation of M on w, with length less or equal to $f(|w|)$. In this search, the machine queries the oracle L' for $\mathcal{O}(\log_2(f(|w|)))$ times, asking, in each of these queries, if a string of the form $1\#w\#\epsilon\#1^k$ and $0\#w\#\epsilon\#1^k$, with $k \leq f(|w|)$, is in L'. Let n_0 be the minimum length of a computation, with length less than or equal to $f(|w|)$. This step is executed deterministically.
2. Next, M' simulates the computations of M, counting how many steps it has already simulated. As soon as a computation has more than n_0 steps, it makes a nondeterministic move, with two possible continuations: one possibility is to accept the input, while the other one is to reject it. The computations with n_0 steps are fully simulated (and the decision of M' in those cases coincide with the decision of M).

It is not hard to see that M' correctly computes the length n_0 of the shortest computation of M on an input word w. It is also clear that the difference between the number of accepting paths and the number of rejecting paths of M' equals the difference between the number of accepting shortest computations and rejecting shortest computations of M. Finally, note that M' works in nondeterministic polynomial time, and it makes $\mathcal{O}(\log n)$ queries to a **NP** language, summed up over all the possible computations. Therefore, if we see M' as a **PP**-machine, it makes exactly the same decision as M, seen as a $\mathbf{PP_{sc}}$-machine.

Clearly, this implies that $\mathbf{PP_{sc}} \subseteq \mathbf{PP^{NP[\log]}_{ctree}}$, and our proof is concluded.

Alternatively, one can see that all the languages from $\mathbf{PP_{sc}}$ can be accepted by deterministic Turing machines working in polynomial time, that are allowed to make $\mathcal{O}(\log n)$ queries to **NP** and exactly one query to **PP**, which gives the decision of the machine, on an input of length n. The only difference from the above idea is that step 2 of the algorithm is replaced by a **PP**-language query.

Another remark is that the idea presented above holds in the case of other classes, like $\oplus\mathbf{P}$ (where $\oplus\mathbf{P}$ is the class of decision problems solvable by a nondeterministic polynomial Turing machine which accepts if and only if the number of accepting paths is even), which was introduced in [4].

One can show, similarly to the above, that $\oplus\mathbf{P}_{sc} \subseteq \oplus\mathbf{P^{NP[\log]}_{ctree}}$ (where $\oplus\mathbf{P^{NP[\log]}_{ctree}}$ is the class of decision problems solvable by a $\oplus\mathbf{P}$-machine which can make a total number of $\mathcal{O}(\log n)$ queries to an **NP**-language in its entire computation tree, on an input of length n). The only difference from the above proof is that in step 2 of the algorithm, as soon as a computation has more than n_0 steps, the machine M makes a nondeterministic move with three possible continuations: two possibilities are to accept the input, and the other is to reject it.

The same idea applies to the class **RP**, of decision problems solvable by a nondeterministic polynomial Turing machine which accepts if and only if at least half of the computation paths accept and rejects if and only if all computation paths reject, introduced in [5]. In this case we get $\mathbf{RP}_{sc} \subseteq \mathbf{RP}^{\mathbf{NP}[\log]}_{\mathbf{ctree}}$ (where $\mathbf{RP}^{\mathbf{NP}[\log]}_{\mathbf{ctree}}$ is the class of decision problems solvable by a **RP**-machine which can make a total number of $\mathcal{O}(\log n)$ queries to an **NP**-language in its entire computation tree, on an input of length n).

According to Remark 2, one can see that the lower bounds $\mathbf{P}^{\mathbf{NP}[\log]} \subseteq \mathbf{PP}_{\mathbf{sc}}$, $\mathbf{P}^{\mathbf{NP}[\log]} \subseteq \oplus\mathbf{P}_{\mathbf{sc}}$ and $\mathbf{P}^{\mathbf{NP}[\log]} \subseteq \mathbf{RP}_{\mathbf{sc}}$ hold. □

Funding: This research was done by the author while being a postdoctoral fellow at the University of Magdeburg, funded by the *Alexander von Humboldt Foundation*, whose support is graciously acknowledged.

Acknowledgments: An extended abstract of this paper was presented at the conference *Computability in Europe 2011: Models of Computation in Context* [21]. We thank to all the reviewers of this paper for their very useful comments and suggestions.

Conflicts of Interest: The author declares no conflict of interest.

References

1. Köbler, J.; Schöning, U.; Wagner, K. The Difference and Truth-Table Hierarchies for NP. *RAIRO Theor. Inform. Appl.* **1987**, *21*, 419–435. [CrossRef]
2. Hemachandra, L. The Strong Exponential Hierarchy Collapses. *J. Comput. Syst. Sci.* **1989**, *39*, 299–322. [CrossRef]
3. Wagner, K.W. Bounded Query Classes. *SIAM J. Comput.* **1990**, *19*, 833–846. [CrossRef]
4. Papadimitriou, C.H.; Zachos, S. Two remarks on the power of counting. In Proceedings of the Theoretical Computer Science, Dortmund, Germany, 5–7 January 1983; Volume 145, pp. 269–276.
5. Gill, J. Computational Complexity of Probabilistic Turing Machines. *SIAM J. Comput.* **1977**, *6*, 675–695. [CrossRef]
6. Beigel, R.; Hemachandra, L.A.; Wechsung, G. Probabilistic Polynomial Time is Closed under Parity Reductions. *Inf. Process. Lett.* **1991**, *37*, 91–94. [CrossRef]
7. Han, Y.; Hemaspaandra, L.; Thierauf, T. Threshold Computation and Cryptographic Security. *SIAM J. Comput.* **1997**, *26*, 59–78. [CrossRef]
8. Papadimitriou, C.M. *Computational Complexity*; Addison-Wesley: Reading, MA, USA, 1994.
9. Manea, F.; Margenstern, M.; Mitrana, V.; Pérez-Jiménez, M.J. A New Characterization of NP, P, and PSPACE with Accepting Hybrid Networks of Evolutionary Processors. *Theory Comput. Syst.* **2010**, *46*, 174–192. [CrossRef]
10. Pérez-Jiménez, M.J. A Computational Complexity Theory in Membrane Computing. In Proceedings of the International Workshop on Membrane Computing, Curtea de Arges, Romania, 24–27 August 2009; Volume 5957, pp. 125–148.
11. Manea, F. Deciding Networks of Evolutionary Processors. In Proceedings of the International Conference on Developments in Language Theory, Milano, Italy, 19–22 July 2011; Volume 6795, pp. 337–349.
12. Manea, F. Complexity results for deciding Networks of Evolutionary Processors. *Theor. Comput. Sci.* **2012**, *456*, 65–79. [CrossRef]
13. Alhazov, A.; Freund, R.; Rogozhin, V.; Rogozhin, Y. Computational completeness of complete, star-like, and linear hybrid networks of evolutionary processors with a small number of processors. *Nat. Comput.* **2016**, *15*, 51–68. [CrossRef]
14. Alhazov, A.; Rogozhin, Y.; Verlan, S. Small Universal Devices. In *Computing with New Resources—Essays Dedicated to Jozef Gruska on the Occasion of His 80th Birthday*; Calude, C.S., Freivalds, R., Iwama, K., Eds.; Springer: Berlin/Heidelberg, Germany, 2014; Volume 8808, pp. 249–263. [CrossRef]
15. Loos, R. On Accepting Networks of Splicing Processors of Size 3. In Proceedings of the Conference on Computability in Europe, Siena, Italy, 18–23 June 2007; Volume 4497, pp. 497–506.
16. Negru, M.C. Networks of polarized splicing processors as problem solvers. *Biosystems* **2019**, *186*, 104037. [CrossRef]
17. Hartmanis, J.; Stearns, R.E. On the Computational Complexity of Algorithms. *Trans. Amer. Math. Soc.* **1965**, *117*, 533–546. [CrossRef]
18. Hopcroft, J.E.; Ullman, J.D. *Introduction to Automata Theory, Languages and Computation*; Addison-Wesley: Reading, MA, USA 1979.
19. Wagner, K.W. More Complicated Questions About Maxima and Minima, and Some Closures of NP. *Theor. Comput. Sci.* **1987**, *51*, 53–80. [CrossRef]
20. Hemaspaandra, E.; Hemaspaandra, L.A.; Rothe, J. Exact Analysis of Dodgson Elections: Lewis Carroll's 1876 Voting System is Complete for Parallel Access to NP. *J. ACM* **1997**, *1256*, 214–224. [CrossRef]
21. Manea, F. Deciding According to the Shortest Computations. In Proceedings of the Conference on Computability in Europe 2011, Sofia, Bulgaria, 27 June–2 July 2011; Volume 6735, pp. 191–200.

axioms

MDPI

Article

The Maximal Complexity of Quasiperiodic Infinite Words

Ludwig Staiger

Institut für Informatik, Martin-Luther-Universität Halle-Wittenberg, D-06099 Halle (Saale), Germany; staiger@informatik.uni-halle.de

Abstract: A quasiperiod of a finite or infinite string is a word whose occurrences cover every part of the string. An infinite string is referred to as quasiperiodic if it has a quasiperiod. We present a characterisation of the set of infinite strings having a certain word q as quasiperiod via a finite language P_q consisting of prefixes of the quasiperiod q. It turns out its star root $\sqrt[*]{P_q}$ is a suffix code having a bounded delay of decipherability. This allows us to calculate the maximal subword (or factor) complexity of quasiperiodic infinite strings having quasiperiod q and further to derive that maximally complex quasiperiodic infinite strings have quasiperiods *aba* or *aabaa*. It is shown that, for every length $l \geq 3$, a word of the form $a^n ba^n$ (or $a^n bba^n$ if l is even) generates the most complex infinite string having this word as quasiperiod. We give the exact ordering of the lengths l with respect to the achievable complexity among all words of length l.

Keywords: quasiperiod; formal language; asymptotic growth; polynomial

MSC: 68Q45

Citation: Staiger, L. The Maximal Complexity of Quasiperiodic Infinite Words. *Axioms* **2021**, *10*, 306. https://doi.org/10.3390/axioms10040306

Academic Editor: Victor Mitrana

Received: 27 September 2021
Accepted: 11 November 2021
Published: 17 November 2021

1. Introduction

In his tutorials [1–3] Solomon Marcus dealt with several properties of infinite words. Among them he considered quasiperiodicity and its influence on measures of symmetry like complexity, recurrence or entropy. One topic of interest was their *subword complexity* (or *factor complexity* [4]). Besides the asymptotic behaviour of the factor complexity, also known as their topological entropy ([4], Section 4.2.2) or [5] Marcus was also interested in the behaviour of the complexity function $f(\xi, n)$ assigning to a natural number $n \in \mathbb{N}$ the number of subwords of the infinite word (ω-word) ξ. Here he was also concerned with recurrences in ω-words and their influence to subword complexity. A well-known fact established by Grillenberger is that the asymptotic subword complexity (or topological entropy) of an almost periodic (or uniformly recurrent) ω-word can be arbitrarily close (but not equal) to the maximal subword complexity (see [4], Theorem 4.4.4).

The present paper summarises results on the subword complexity of infinite words obtained in [6–8]. We study in detail the structure of the set of infinite words having a certain word q as quasiperiod and how this is connected with the set of finite words with the same quasiperiod. Moreover, we address a question raised in [9] about the maximally achievable subword complexity of a quasiperiodic infinite word.

A first result shows that for every word q there is a value $\lambda_q, 1 \leq \lambda_q < 2$, such that, for every infinite word ξ with quasiperiod q, the complexity function $f(\xi, n)$ is bounded by $O(1) \cdot \lambda_q^n$, and this bound is achieved for certain infinite words having quasiperiod q. The maximally possible value for λ_q is $\lambda_q = t_P \approx 1.324718$, where t_P is the smallest Pisot-Vijayaraghavan number, that is, the unique real root t_P of the cubic polynomial $x^3 - x - 1$.

As a generalisation of the above-mentioned questions [2,9] we estimate, for every length $n \geq 3$, the values $\lambda_n = \max\{\lambda_q : |q| = n\}$, their ordering and the words $q, |q| = n$, for which $\lambda_q = \lambda_n$. It appears that a two letter alphabet is sufficient for achieving the maximal complexity λ_n.

In order to prove these properties we start with a general investigation of quasiperiodicity of words (as e.g., in [10–12]) and infinite words.

The paper is organised as follows. After introducing some notation we derive in Section 3 a characterisation of quasiperiodic words and ω-words having a certain quasiperiod q. Moreover, we use the finite basis sets P_q and its dual R_q ($\mathcal{L}(q)$ and $\mathcal{R}(q)$ in [12]) from which the sets of quasiperiodic words or ω-words having quasiperiod q can be constructed. In Section 4 it is then proved that the star root of P_q is a suffix code having a bounded delay of decipherability and, dually, the star root of R_q is a prefix code.

This much prerequisites allow us, in Section 5, to estimate the number of subwords of the language Q_q of all quasiperiodic words having quasiperiod q. It turns out that $c_{q,1} \cdot \lambda_q^n \leq f(Q_q, n) \leq c_{q,2} \cdot \lambda_q^n$ where $f(Q_q, n)$ is the number of subwords of length n of words in Q_q and $1 \leq \lambda_q \leq t_P$ depends on q. We construct, for every quasiperiod q, a quasiperiodic ω-word ξ_q with quasiperiod q whose subword complexity $f(\xi_q, n)$ is maximal.

The values λ_q turn out to be maximal positive roots of polynomials associated with the star root $\sqrt[*]{P_q}$. Section 6 deals with the properties of those polynomials. This allows to compare the roots λ_q.

The following Sections 7 and 8 deal with the proof of the above mentioned results on the values λ_q and $\lambda_n = \max\{\lambda_q : |q| = n\}$. Here we derive also the complete ordering of the values λ_n.

2. Notation and Preliminaries

In this section we introduce the notation used throughout the paper. By $\mathbb{N} = \{0, 1, 2, \ldots\}$ we denote the set of natural numbers. Let X be an alphabet of cardinality $|X| = r \geq 2$, and let throughout the paper $a, b \in X, a \neq b$, be two different letters. By X^* we denote the set of finite words on X, including the *empty word* e, and X^ω is the set of infinite strings (ω-words) over X. Subsets of X^* will be referred to as *languages* and subsets of X^ω as *ω-languages*.

For $w \in X^*$ and $\eta \in X^* \cup X^\omega$ let $w \cdot \eta$ be their *concatenation*. This concatenation product extends in an obvious way to subsets $L \subseteq X^*$ and $B \subseteq X^* \cup X^\omega$. For a language L let $L^* := \bigcup_{i \in \mathbb{N}} L^i$, and by $L^\omega := \{w_1 \cdots w_i \cdots : w_i \in L \setminus \{e\}\}$ we denote the set of infinite strings formed by concatenating words in L. The smallest subset of a language L which generates L^* is called its *star root* $\sqrt[*]{L}$ [13]. It holds

$$\sqrt[*]{L} = (L \setminus \{e\}) \setminus (L \setminus \{e\})^2 \cdot L^* .$$

Furthermore $|w|$ is the *length* of the word $w \in X^*$ and $\mathbf{pref}(B)$ is the set of all finite prefixes of the strings in $B \subseteq X^* \cup X^\omega$. We shall abbreviate $w \in \mathbf{pref}(\eta)$ ($\eta \in X^* \cup X^\omega$) by $w \sqsubseteq \eta$.

We denote by $B/w := \{\eta : w \cdot \eta \in B\}$ the *left derivative* of the set $B \subseteq X^* \cup X^\omega$. As usual, a language $L \subseteq X^*$ is *regular* provided it is accepted by a finite automaton. An equivalent condition is that its set of left derivatives $\{L/w : w \in X^*\}$ is finite.

The sets of infixes of B or η are $\mathbf{infix}(B) := \bigcup_{w \in X^*} \mathbf{pref}(B/w)$ and $\mathbf{infix}(\eta) := \bigcup_{w \in X^*} \mathbf{pref}(\{\eta\}/w)$, respectively. In the sequel we assume the reader to be familiar with basic facts of language theory.

We call a word $w \in X^* \setminus \{e\}$ *primitive* if $w = v^n$ implies $n = 1$, that is, w is not the power of a shorter word, and we call $w \in X^* \setminus \{e\}$ *overlap-free* if none of its proper prefixes is a suffix of w. The following facts are known (e.g., [14,15]).

Fact 1. *Every word $w \in X^* \setminus \{e\}$ has a unique representation $w = v^n$ where v is primitive.*

Fact 2. *Let $q, v, w \in X^*, 0 < |v| < |q|$. If $v \cdot q = q \cdot w$ then $v = u \cdot u'$, $q = (u \cdot u')^\kappa \cdot u$ and $w = u' \cdot u$ for some $u, u' \in X^*, u \neq e$, and $\kappa \in \mathbb{N}$. In particular, q is not overlap-free.*

Fact 3. *If $w \cdot v = v \cdot w$, $w, v \in X^*$ then w, v are powers of a common (primitive) word.*

As usual a language $L \subseteq X^*$ is called a *code* provided $w_1 \cdots w_l = v_1 \cdots v_k$ for $w_1, \ldots, w_l, v_1, \ldots, v_k \in L$ implies $l = k$ and $w_i = v_i$. A code L is said to be a *prefix code* (*suffix code*) provided no codeword is a prefix (suffix) of another codeword.

3. Quasiperiodicity

3.1. General Properties

The notion of quasiperiodicity can be formalised in the following manner. A finite or infinite word $\eta \in X^* \cup X^\omega$ is referred to as *quasiperiodic* with quasiperiod $q \in X^* \setminus \{e\}$ provided that for every $j < |\eta| \in \mathbb{N} \cup \{\infty\}$ there is a prefix $u_j \sqsubseteq \eta$ of length $j - |q| < |u_j| \leq j$ such that $u_j \cdot q \sqsubseteq \eta$, that is, for every $w \sqsubseteq \eta$ the relation $u_{|w|} \sqsubset w \sqsubseteq u_{|w|} \cdot q$ is valid. Informally, η has quasiperiod q if every position of η occurs within some occurrence of q in η [11,12].

Let for $q \in X^* \setminus \{e\}$, Q_q be the set of quasiperiodic words with quasiperiod q. Then $\{q\}^* \subseteq Q_q = Q_q^*$ and $Q_q \setminus \{e\} \subseteq X^* \cdot q \cap q \cdot X^*$. In order to describe the set of quasiperiodic strings having a certain quasiperiod $q \in X^* \setminus \{e\}$ the following definition is helpful.

Definition 1. *A family $(w_i)_{i=1}^{\ell}$, $\ell \in \mathbb{N} \cup \{\infty\}$, of words $w_i \in X^* \cdot q$ is referred to as a q-chain provided $w_1 = q$, $w_i \sqsubset w_{i+1}$ and $|w_{i+1}| - |w_i| \leq |q|$.*

It holds the following.

Lemma 1.

1. *$w \in Q_q \setminus \{e\}$ if and only if there is a q-chain $(w_i)_{i=1}^{\ell}$ such that $w_\ell = w$.*
2. *An ω-word $\xi \in X^\omega$ is quasiperiodic with quasiperiod q if and only if there is a q-chain $(w_i)_{i=1}^{\infty}$ such that $w_i \sqsubset \xi$.*

Proof. It suffices to show how a family $(u_j)_{j=0}^{|\eta|-1}$ can be converted to a q-chain $(w_i)_{i=1}^{\ell}$ and vice versa.

Consider $\eta \in X^* \cup X^\omega$ and let $(u_j)_{j=0}^{|\eta|-1}$ be a family such that $u_j \cdot q \sqsubseteq \eta$ and $j - |q| < |u_j| \leq j$ for $j < |\eta|$.

Define $w_1 := q$ and $w_{i+1} := u_{|w_i|} \cdot q$ as long as $|w_i| < |\eta|$. Then $w_i \sqsubseteq \eta$ and $|w_i| < |w_{i+1}| = |u_{|w_i|} \cdot q| \leq |w_i| + |q|$. Thus $(w_i)_{i=1}^{\ell}$ is a q-chain with $w_i \sqsubseteq \eta$.

Conversely, let $(w_i)_{i=1}^{\ell}$ be a q-chain such that $w_i \sqsubseteq \eta$ and set

$$u_j := \max\nolimits_{\sqsubseteq}\{w' : \exists i(w' \cdot q = w_i \wedge |w'| \leq j)\}, \text{ for } j < |\eta|.$$

By definition, $u_j \cdot q \sqsubseteq \eta$ and $|u_j| \leq j$. Assume $|u_j| \leq j - |q|$ and $u_j \cdot q = w_i$. Then $|w_i| \leq j < |\eta|$. Consequently, in the q-chain there is a successor w_{i+1}, $|w_{i+1}| \leq |w_i| + |q| \leq j + |q|$. Let $w_{i+1} = w'' \cdot q$. Then $u_j \sqsubset w''$ and $|w''| \leq j$ which contradicts the maximality of u_j. □

Lemma 1 yields the following consequences.

Corollary 1. *Let $u \in \mathbf{pref}(Q_q)$. Then there are words $w, w' \in Q_q$ such that $w \sqsubseteq u \sqsubseteq w'$ and $|u| - |w|, |w'| - |u| \leq |q|$.*

Corollary 2. *Let $\xi \in X^\omega$. Then the following are equivalent.*

1. *ξ is quasiperiodic with quasiperiod q.*
2. *$\mathbf{pref}(\xi) \cap Q_q$ is infinite.*
3. *$\mathbf{pref}(\xi) \subseteq \mathbf{pref}(Q_q)$.*

3.2. Finite Generators for Quasiperiodic Words

In this part we consider the finite languages P_q and R_q ($\mathcal{L}(q)$ and $\mathcal{R}(q)$ in [12]) which generate the set of quasiperiodic words as well as the set of quasiperiodic ω-words having quasiperiod q.

We set

$$P_q := \{v : e \sqsubset v \sqsubseteq q \sqsubset v \cdot q\} = \{v : \exists v'(v' \sqsubset q \wedge v \cdot v' = q)\}. \quad (1)$$

Then we have the following properties.

Proposition 1.

1. *$q \in P_q$ and $P_q = \{q\}$ if and only if q is overlap-free.*
2. $Q_q = P_q^* \cdot q \cup \{e\} \subseteq P_q^*$
3. $\mathbf{pref}(Q_q) = \mathbf{pref}(P_q^*) = P_q^* \cdot \mathbf{pref}(q)$

Proof. 1. $q \in P_q$ is obvious and and the equivalence follows immediately from the definition of P_q.

2. In order to prove $Q_q \subseteq P_q^* \cdot q \cup \{e\}$ we show that $w_i \in P_q^* \cdot q$ for every q-chain $(w_i)_{i=1}^{\ell}$. This is certainly true for $w_1 = q$. Now proceed by induction on i. Let $w_i = w_i' \cdot q \in P_q^* \cdot q$ and $w_{i+1} = w_{i+1}' \cdot q$. Then $w_i' \cdot v_i = w_{i+1}'$. Now from $w_i \sqsubset w_{i+1}$ we obtain $e \sqsubset v_i \sqsubseteq q \sqsubset v_i \cdot q$, that is, $v_i \in P_q$.

Conversely, let $v_i \in P_q$ and consider $v_1 \cdots v_\ell \cdot q$. Since $q \sqsubseteq v_i \cdot q$ the family $(v_1 \cdots v_j \cdot q)_{j=0}^{\ell}$ is a q-chain. This shows $P_q^* \cdot q \cup \{e\} \subseteq Q_q$.

3. is an immediate consequence of 2. □

Proposition 1 and Corollary 2 imply the following characterisation of ω-words having quasiperiod q.

$$\{\xi : \xi \in X^\omega \wedge \xi \text{ has quasiperiod } q\} = P_q^\omega \quad (2)$$

Proof. Since P_q is finite, $P_q^\omega = \{\xi : \xi \in X^\omega \wedge \mathbf{pref}(\xi) \subseteq \mathbf{pref}(P_q^*)\}$. □

A dual generator of Q_q is obtained by the right-to-left duality of reading words using the suffix relation $\leq_s$ instead of the prefix relation $\sqsubseteq$.

$$R_q := \{v : e <_s v \leq_s q <_s q \cdot v\} = \{v : \exists v'(v' <_s q \wedge v' \cdot v = q)\}. \quad (3)$$

Analogously to Proposition 1 we obtain

Proposition 2.

1. *$q \in R_q$ and $R_q = \{q\}$ if and only if q is overlap-free.*
2. *$Q_q = q \cdot R_q^* \cup \{e\} \subseteq R_q^*$, and*
3. *$\mathbf{pref}(Q_q) = \mathbf{pref}(q) \cup q \cdot \mathbf{pref}(R_q^*)$.*

The proof of Items 1 and 2 is similar to the proof of Proposition 1 using the reversed version of q-chain, and Item 3 then follows from Item 2. A slight difference appears with an analogy to Equation (2).

$$\{\xi : \xi \in X^\omega \wedge \xi \text{ has quasiperiod } q\} = q \cdot R_q^\omega \subseteq R_q^\omega \quad (4)$$

Here the last inclusion might be proper, e.g., for $q = aba$ where $R_{aba}^\omega = \{ba, aba\}^\omega \neq aba \cdot R_{aba}^\omega$.

An alternative derivation of the languages P_q and R_q can be found in Definition 2 of [12]. Here the borders, that is, prefixes which are simultaneously suffixes of the quasiperiod q, are used:

$$\begin{aligned} P_q &= \{v : \exists w(w \sqsubset q \wedge w <_s q \wedge q = v \cdot w)\}\,, \text{ and} \\ R_q &= \{v : \exists w(w \sqsubset q \wedge w <_s q \wedge q = w \cdot v)\}\,. \end{aligned}$$

In the subsequent sections we focus on the investigation of P_q due to the left-to-right direction of ω-words.

3.3. Combinatorial Properties of P_q

We investigate basic properties of P_q using simple facts from combinatorics on words (see e.g., [14–16]).

Proposition 3. *$v \in P_q$ if and only if $|v| \leq |q|$ and there is a prefix $\bar{v} \sqsubset v$ such that $q = v^k \cdot \bar{v}$ for $k = \lfloor |q|/|v| \rfloor$.*

This is an immediate consequence of Fact 2.

Corollary 3. *$v \in P_q$ if and only if $|v| \leq |q|$ and there is a $k' \in \mathbb{N}$ such that $q \sqsubseteq v^{k'}$.*

Now set $q_0 := \min_{\sqsubseteq} P_q$. Then in view of Proposition 3 and Corollary 3 we have the following canonical representation.

$$q = q_0^k \cdot \bar{q} \text{ where } k = \lfloor |q|/|q_0| \rfloor \text{ and } \bar{q} \sqsubset q_0\,. \tag{5}$$

We will refer to q_0 as the *repeated prefix* and to k as the *repetition factor*. If $|q_0| > |q|/2$, that is, if $k = 1$ we will refer to q as *irreducible*. (Reducible words are also known as periodic words [10,11].)

Corollary 4. *Every word $v \in \sqrt[*]{P_q}$ is primitive.*

Proof. Assume $v = v_1^l$ for some $v \in \sqrt[*]{P_q}$ and $l > 1$. Then $q \sqsubseteq v^{k'} = v_1^{l \cdot k'}$, and, according to Corollary 3 $v_1 \in P_q$ contradicting $v \in \sqrt[*]{P_q}$. □

Proposition 4. *Let $q \in X^*, q \neq e$, $q_0 = \min_{\sqsubseteq} P_q$, $q = q_0^k \cdot \bar{q}$ and $v \in P_q^* \setminus \{e\}$.*

1. *If $w \sqsubseteq q$ then $v \cdot w \sqsubseteq q$ or $q \sqsubseteq v \cdot w$.*
2. *If $w \cdot v \sqsubseteq q$ then $w \in \{q_0\}^*$.*

Proof. From Proposition 1.2 we know $v \cdot q \in P_q^* \cdot q \subseteq Q_q \subseteq q \cdot X^*$. Consequently, $q \sqsubseteq v \cdot q$. Then $v \cdot w \sqsubseteq v \cdot q$ implies $v \cdot w \sqsubseteq q$ or $q \sqsubseteq v \cdot w$ according to whether $|w \cdot w| \leq |q|$ or not.

Since $q_0 \sqsubseteq v$, it suffices to prove the second assertion for q_0. First one observes that, $w \sqsubseteq q$ and $|w| \leq |q| - |q_0|$. Thus $w \sqsubseteq q_0^{k-1} \cdot \bar{q}$. Therefore, we have $w \cdot q_0 \sqsubseteq q$ and $q_0 \cdot w \sqsubseteq q$ which implies $w \cdot q_0 = q_0 \cdot w$ and, according to Fact 3, w and q_0 are powers of a common word. The assertion follows because q_0 is primitive. □

Next we derive a lower bound on the lengths of words in $P_q \setminus \{q_0\}^*$.

To this end, we use the Theorem of Fine and Wilf.

Theorem 1 ([17]). *Let $v, w \in X^*$. Suppose v^m and w^n, for some $m, n \in \mathbb{N}$, have a common prefix of length $|v| + |w| - \gcd(|v|, |w|)$. Then v and w are powers of a common word $u \in X^*$ of length $|u| = \gcd(|v|, |w|)$. (Here $\gcd(k, l)$ denotes the greatest common divisor of two numbers $k, l \in \mathbb{N}$.)*

Proposition 5. *Let $q \in X^*, q \neq e$, $q_0 = \min_{\sqsubseteq} P_q$, $q = q_0^k \cdot \bar{q}$ and $v \in P_q \setminus \{q_0\}^*$. Then $|v| > |q| - |q_0| + \gcd(|v|, |q_0|)$.*

Proof. If $q_0, v \in P_q$ Corollary 3 and Equation (5) imply that q is a common prefix of q_0^{k+1} and $v^{k'}$ for some $k' \in \mathbb{N}$. If $|v| \leq |q| - |q_0| + \gcd(|v|, |q_0|)$ then by Theorem 1 q_0 and v are powers of a common word, that is, v is a power of the primitive word q_0. □

Corollary 5. $\sqrt[*]{P_q} = P_q \setminus q_0^2 \cdot \{q_0\}^*$

Proof. It suffices to show $P_q \cap P_q^2 \cdot P_q^* \subseteq \{q_0\}^*$. To this end observe that in view of Proposition 5 $|v \cdot v'| > |q|$ whenever $v \in P_q \setminus \{q_0\}^*$ or $v' \in P_q \setminus \{q_0\}^*$. □

As an immediate consequence we obtain that $\sqrt[*]{P_q} = P_q$ if and only if q is an irreducible quasiperiod. Moreover, Proposition 5 shows that

$$\sqrt[*]{P_q} \subseteq \{q_0\} \cup \{v' : v' \sqsubseteq q \wedge |v'| > |q| - |q_0| + \gcd(|v'|, |q_0|)\}. \tag{6}$$

3.4. The Reduced Quasiperiod $\hat{q}$

Next we investigate the relation between a quasiperiod $q = q_0^k \cdot \bar{q}$ where $q_0 = \min_{\sqsubseteq} P_q$ and $\bar{q} \sqsubset q_0$ and its *reduced quasiperiod* $\hat{q} := q_0 \cdot \bar{q}$. Since $q \in Q_{\hat{q}}$, we have $Q_{\hat{q}} \supseteq Q_q$.

We continue with a relation between P_q and $P_{\hat{q}}$. It is obvious that $q_0^i \in P_q$ for every $i = 1, \ldots, k$ and

$$P_{\hat{q}} \subseteq \{v : \hat{q}_0 \sqsubseteq v \sqsubseteq \hat{q}\}. \tag{7}$$

Lemma 2 ([7], Lemma 2.2). *Let $q \in X^*, q \neq e$, $q_0 = \min_{\sqsubseteq} P_q$, $q = q_0^k \cdot \bar{q}$ and $\hat{q} = q_0 \cdot \bar{q}$ the reduced quasiperiod of q. Then*

$$P_q = \{q_0^i : i = 1, \ldots, k-1\} \cup \{q_0^{k-1} \cdot v : v \in P_{\hat{q}}\}.$$

Proof. Consider $v \in P_{\hat{q}}$. Then $v \sqsubseteq q_0\bar{q} \sqsubset v \cdot q_0\bar{q}$, and, consequently, $q_0^{k-1} \cdot v \sqsubseteq q_0^k \cdot \bar{q} \sqsubset q_0^{k-1} \cdot v \cdot q_0\bar{q} \sqsubset q_0^{k-1} \cdot v \cdot q_0^k \cdot \bar{q}$, that is, $q_0^{k-1} \cdot v \in P_q$.

Conversely, let $v' \in P_q$ and $v' \notin \{q_0^i : i = 1, \ldots, k-1\}$. Then, according to Proposition 5 there is a unique $v \neq e$ such that $v' = q_0^{k-1} \cdot v$. Now $v' = q_0^{k-1} \cdot v \sqsubseteq q = q_0^k \cdot \bar{q} \sqsubset v' \cdot q = q_0^{k-1} \cdot v \cdot q_0^k \cdot \bar{q}$ implies $v \sqsubseteq q_0 \cdot \bar{q} \sqsubset v \cdot q_0^k \cdot \bar{q}$. Since $|v| \leq |q_0 \cdot \bar{q}|$ and $q_0 \cdot \bar{q} \sqsubseteq q_0^k \cdot \bar{q}$, we have $v \sqsubseteq q_0 \cdot \bar{q} \sqsubset v \cdot q_0 \cdot \bar{q}$. □

Together with Corollary 5 this implies

$$P_q \setminus \{q_0\}^* = \sqrt[*]{P_q} \setminus \{q_0\}^* = q_0^{k-1} \cdot (P_{\hat{q}} \setminus \{q_0\}). \tag{8}$$

Moreover, we have the following.

Corollary 6. $|\sqrt[*]{P_q}| = 1$ *if and only if $q \in \{q_0\}^*$ and q_0 is overlap-free.*

Proof. Since $q_0 \in \sqrt[*]{P_q}$, $|\sqrt[*]{P_q}| = 1$ is equivalent with $\sqrt[*]{P_q} = \{q_0\}$ or, according to Equation (8), with $P_{\hat{q}} = \{q_0\}$. This amounts to $\hat{q} = q_0$ and, following Proposition 1.1 $\hat{q} = q_0$ has to be overlap-free. □

For the repeated prefix $\hat{q}_0$ of $\hat{q}$ we have the obvious relation $|\hat{q}_0| > |\bar{q}|$. In case $\hat{q}_0 \neq q_0$ we can improve this.

Lemma 3. *Let $q = q_0^k \cdot \bar{q}$ with $k \geq 2$, $\bar{q} \sqsubset q_0$ and $\hat{q} = q_0 \cdot \bar{q}$. If $\hat{q}_0 \neq q_0$ then*

$$\bar{q} \sqsubset \hat{q}_0 \sqsubset q_0 \text{ and } |\hat{q}_0| > |\bar{q}| + \gcd(|q_0|, |\hat{q}_0|),$$

and there is a nonempty suffix $v \neq e$ of q_0 such that $v \sqsubset \hat{q}_0$ and $v \cdot \bar{q} \sqsubset \hat{q}_0^2$.

Proof. We have $\bar{q} \sqsubseteq q_0$ and, since $q_0 \in P_{\hat{q}}$, also $\hat{q}_0 \sqsubseteq q_0$. Moreover, $\hat{q} \sqsubseteq q_0^2$ and $\hat{q} \sqsubseteq \hat{q}_0^{k'}$ for some $k' \in \mathbb{N}$. Since $q_0 \neq \hat{q}_0$ and both prefixes are primitive words, in view of Theorem 1 as a common prefix of q_0^2 and $\hat{q}_0^{|q_0|}$ the word $\hat{q} = q_0 \cdot \bar{q}$ has to satisfy $|\hat{q}| < |q_0| + |\hat{q}_0| - \gcd(|q_0|, |\hat{q}_0|)$, that is, $|\hat{q}_0| > |\bar{q}| + \gcd(|q_0|, |\hat{q}_0|)$. The assertion $\bar{q} \sqsubset \hat{q}_0 \sqsubset q_0$ now follows from a comparison of the lengths of $\bar{q}, \hat{q}_0 \sqsubseteq q_0$.

Now, let v be the suffix of q_0 defined by $\hat{q}_0^{k'} \cdot v = q_0 \sqsubset \hat{q}_0^{k'+1}$. Then $v \sqsubset \hat{q}_0$ and $v \cdot \bar{q} \sqsubset (\hat{q}_0)^2$. □

3.5. Primitivity and Superprimitivity

In this section we consider the inclusion relations between the languages $Q_q, q \neq e$. Analogously to the primitivity of words in [10–12] a word was referred to as *superprimitive* if it is not covered by a shorter one. This leads to the following definition.

Definition 2 (superprimitive). *A non-empty word $q \in X^* \setminus \{e\}$ is* superprimitive *if and only if Q_q is maximal w.r.t. "$\subseteq$" in the family $\{Q_q : q \in X^* \setminus \{e\}\}$.*

The next proposition relates the irreducibility of quasiperiods to superprimitivity.

Proposition 6 ([12], Remark 4). *If $q \in X^* \setminus \{e\}$ is superprimitive then $|\min_{\sqsubseteq} P_q| > |q|/2$, and if $|\min_{\sqsubseteq} P_q| > |q|/2$ then q is primitive.*

Proof. If $q_0 = \min_{\sqsubseteq} P_q$ and $|q_0| \leq |q|/2$ then $q = q_0^k \cdot \bar{q}$ for some $\bar{q} \sqsubset q_0$. Thus $q \in Q_{q_0\bar{q}}$ and $q_0\bar{q} \notin Q_q$.

As $q = q'^m$ with $m > 1$ implies $|q_0| \leq |q'| \leq |q|/2$, the other assertion follows. □

The converse of Proposition 6 is not valid.

Example 1. *Let $q = abaabaababaab$. Then $P_q = \{abaabaab, abaabaababa, q\}$, and $|\min_{\sqsubseteq} P_q| = 8 > 13/2$ but as $abaabaababaab \in Q_{abaab}$ the word q is not superprimitive.*

The word $q = ababa$ is primitive but $q_0 = ab$ has $|q_0| \leq |q|/2$.

In contrast to the fact that the word $q_0 = \min_{\sqsubseteq} P_q$ is always primitive, it need not satisfy $|\min_{\sqsubseteq} P_{q_0}| > |q_0|/2$ let alone be superprimitive..

Example 2. *$q = aabaaabaaaa$ has $q_0 = aabaaabaa$ which, in turn has $P_{q_0} = \{aaba, aabaaaba, q_0\}$ with $|aaba| = 4 < |q_0|/2$.*

It turns out that every language Q_v is contained in a unique maximal Q_q. To this end we derive the following lemma (cf. also [10,11]).

Lemma 4. *Let $v \in Q_q$ and $u \in$ **infix**$(v) \cap q \cdot X^* \cap X^* \cdot q$. Then $u \in Q_q$.*

For the sake of completeness we give a proof.

Proof. We use a maximal q-chain $(w_i)_{i=1}^n$ with $w_n = v$. Assume $v = u_1 \cdot u \cdot u_2$. Since u has q as prefix and suffix, there are $1 \leq j \leq l \leq n$ such that $w_j = u_1 \cdot q$ and $w_l = u_1 \cdot u$. Let, for $1 \leq i \leq l - j + 1$, the words w'_i be defined by $w_{i+j-1} = u_1 \cdot w'_i$. Then $(w'_i)_{i=1}^{l-j+1}$ is a q-chain with $w_{l-j+1} = u$, that is, $u \in Q_q$. □

Corollary 7. *If $v \in Q_q \cap Q_u$ and $|q| < |u|$ then $Q_u \subseteq Q_q$.*

The corollary shows that every language Q_v is contained in a unique maximal Q_q and that two languages Q_u, Q_q are either disjoint or compatible w.r.t. set inclusion. The latter is not true for ω-languages.

Example 3. *Let $q = aabaa$ and $u = aabaaa$. Then $q^\omega \notin P_u^\omega$, $u^\omega \notin P_q^\omega$ but $P_u^\omega \cap P_q^\omega \supseteq aa \cdot \{baaa, baaaa\}^\omega$.*

4. P_q and R_q as Codes

In this section we investigate in more detail the properties of the star root of P_q. It turns out that $\sqrt[*]{P_q}$ is a suffix code which, additionally, has a bounded delay of decipherability. This delay is closely related to the largest power of q_0 being a prefix of q.

According to [14,18–20] a subset $C \subseteq X^*$ is a code of a *delay of decipherability* $m \in \mathbb{N}$ if and only if for all $v, v', w_1, \ldots, w_m \in C$ and $u \in C^*$ the relation $v \cdot w_1 \cdots w_m \sqsubseteq v' \cdot u$ implies $v = v'$. Observe that $C \subseteq X^*$ is a prefix code if and only if C has delay 0.

First we show that $\sqrt[*]{P_q}$ is a suffix code. This generalises Proposition 7 of [12].

Proposition 7. *$\sqrt[*]{P_q}$ is a suffix code, and $\sqrt[*]{R_q}$ is a prefix code.*

Proof. Assume $u = w \cdot v$ for some $u, v \in \sqrt[*]{P_q}$, $u \neq v$. Then $u \sqsubseteq q$ and Proposition 4 (2) proves $w \in \{q_0\}^* \setminus \{e\}$. Consequently, $|v| \leq |q| - |q_0|$. Now Proposition 5 implies $v \in \{q_0\}^*$ and hence $u \in \{q_0\}^*$. Since $u, v \in \sqrt[*]{P_q}$, we obtain $u = v = q_0$ contradicting $u \neq v$.

Using the duality of P_q and R_q one shows in an analogous manner that $\sqrt[*]{R_q}$ is a prefix code. □

An easy consequence of Proposition 7 is the Left and Right Normal Form of a quasiperiodic string ([12], Proposition 8).

Corollary 8 (Normal Form). *Every word $w \in Q_q$ has a unique factorisation $w = v_1 \cdot v_2 \cdots v_n$ into words $v_i \in \sqrt[*]{P_q}$ ($\sqrt[*]{R_q}$, respectively).*

Since $\sqrt[*]{R_q}$ is a prefix code while the words $v \in P_q$ are prefixes of each other, we obtain $|\sqrt[*]{P_q} \cap \sqrt[*]{R_q}| = 1$ generalising Remark 5 of [12]. In fact $\sqrt[*]{P_q} \cap \sqrt[*]{R_q} = \{q\}$ or $\sqrt[*]{P_q} \cap \sqrt[*]{R_q} = \{q_0\}$ depending on whether $q \neq q_0^k$ or not.

We continue this part by investigating the delay of decipherability of $\sqrt[*]{P_q}$. We prove that the delay depends on the repetition factor k.

Theorem 2. *Let $q \in X^* \setminus \{e\}$, $q_0 = \min_{\sqsubseteq} P_q$, and $|\sqrt[*]{P_q}| > 1$. Then $\sqrt[*]{P_q}$ is a code having a delay of decipherability of k or $k+1$.*

Proof. If $|\sqrt[*]{P_q}| > 1$ then in view of Proposition 5 there is a $q' \in \sqrt[*]{P_q}$ with $|q'| > |q| - |q_0|$. Since $q' \in P_q$, we have $q \sqsubseteq q' \cdot q_0 \sqsubseteq q' \cdot q$. Consequently, $q_0 \cdot q_0^{k-1} \sqsubseteq q \sqsubseteq q' \cdot q_0$, that is, the delay of decipherability is at least k.

To prove the converse we show that for $q \sqsubseteq q_0^m$ the delay cannot exceed m.

Assume the contrary, that is, $v \cdot w_1 \cdots w_{m+1} \sqsubseteq v' \cdot u$ for some words $v, v', w_1, \ldots, w_{m+1} \in \sqrt[*]{P_q}$, $v \neq v'$, and $u \in P_q^*$. From Proposition 4 (1) we obtain $u \sqsubseteq q$ or $q \sqsubseteq u$ and, since $|w_i| \geq |q_0|$, also $q \sqsubseteq w_1 \cdots w_{m+1}$.

If $v \sqsubset v'$, in view of the inequality $|v| + |q| \geq |v'| + |q_0|$ our assumption yields $v' \cdot q_0 \sqsubseteq v \cdot q$. Therefore, $w \cdot q_0 \sqsubseteq q$ for the word $w \neq e$ with $v \cdot w = v'$ and, according to Proposition 4 (2) $w \in \{q_0\}^*$. This contradicts the fact that $\sqrt[*]{P_q}$ is a suffix code.

If $v' \sqsubset v$, then $|u| > |w_1 \cdots w_{m+1}| \geq |q|$, and via $|v'| + |q| \geq |v| + |q_0|$ we obtain $v \cdot q_0 \sqsubseteq v' \cdot q$ from our assumption. This yields the same contradiction as in the case $v \sqsubset v'$.

The observation $q \sqsubseteq q_0^{k+1}$ finishes the proof. □

For $q = q_0^k$ the preceding proof shows the following.

Corollary 9. *If $q = q_0^k$ and $|\sqrt[*]{P_q}| > 1$ then $\sqrt[*]{P_q}$ has a delay of decipherability of exactly k.*

Thus, if $|\sqrt[*]{P_q}| > 1$ and $q \neq q_0^k$ the code $\sqrt[*]{P_q}$ may have a minimum delay of decipherability of k or $k+1$. We provide examples that both cases are possible.

Example 4. *Let $q :=$ aabaaaaba. Then $q_0 =$ aabaa, $k = 1$ and $\sqrt[*]{P_q} = P_q = \{ q_0, \textit{aabaaaab}, q \}$ which is a code having a delay of decipherability 2.*

$$\begin{array}{llllll} \textit{Indeed} & aabaaaabaa & = & q_0 \cdot q_0 & \sqsubseteq & q \cdot q_0 \quad \textit{or} \\ & aabaaaabaa & = & q_0 \cdot q_0 & \sqsubseteq & aabaaaab \cdot q_0 \,. \end{array}$$

Moreover, in Example 4, $q \cdot q_0 \notin Q_q$. Thus our example shows also that $q \cdot P_q^*$ need not be contained in Q_q.

Example 5. *Let $q := aba$. Then $k = 1$ and $P_q = \{ab, aba\}$ is a code having a delay of decipherability* 1. □

Since $\sqrt[*]{R_q}$ is a prefix code, every ω-word $\xi \in R_q^\omega$ has a unique factorisation into words $w \in \sqrt[*]{R_q}$. For suffix codes the situation is, in general, different. Consider e.g., the suffix code $\{b, ba, aa\}$. Property 4 (ii) of [20] (see also ([21], Proposition 1.9)) shows that codes of bounded delay of decipherability also admit a unique factorisation of ω-words. Thus we obtain from Theorem 2.

Lemma 5 (Normal Form for quasiperiodic ω-words)**.** *Every ω-word $\xi \in P_q^\omega$ has a unique factorisation $\xi = v_1 \cdot v_2 \cdots v_i \cdots$ into words $v_i \in \sqrt[*]{P_q}$.*

5. Subword Complexity

In this section we investigate upper bounds on the the subword complexity function $f(\xi, n)$ for quasiperiodic ω-words. If $\xi \in X^\omega$ is quasiperiodic with quasiperiod q then Proposition 3 and Corollary 3 show $\mathbf{infix}(\xi) \subseteq \mathbf{infix}(P_q^*)$. Thus

$$f(\xi, n) \le |\mathbf{infix}(P_q^*) \cap X^n| \text{ for } \xi \in P_q^\omega \,. \tag{9}$$

Similar to ([22], Proposition 5.5) let $\xi_q := \prod_{v \in P_q^* \setminus \{e\}} v$. This implies $\mathbf{infix}(\xi_q) = \mathbf{infix}(P_q^*)$. Consequently, the tight upper bound on the subword complexity of quasiperiodic ω-words having a certain quasiperiod q is $f_q(n) := f(\xi_q, n) = |\mathbf{infix}(P_q^*) \cap X^n|$. Observe that in view of Propositions 1 and 2 the identity

$$\mathbf{infix}(P_q^*) = \mathbf{infix}(R_q^*) = \mathbf{infix}(Q_q) \tag{10}$$

holds.

The asymptotic upper bound on the subword complexity $f_q(n)$ is obtained from

$$\lambda_q = \limsup_{n \to \infty} \sqrt[n]{|\mathbf{infix}(P_q^*) \cap X^n|}, \tag{11}$$

that is, for large n, $f_q(n) \le \hat{\lambda}^n$ whenever $\hat{\lambda} > \lambda_q$.

The following facts are known from the theory of formal power series (cf. [23,24]). As $\mathbf{infix}(P_q^*)$ is a regular language the power series $\sum_{n \in \mathbb{N}} f_q(n) \cdot t^n$ is a rational series and, therefore, f_q satisfies a recurrence relation

$$f_q(n + k) = \sum_{i=0}^{k-1} a_i \cdot f_q(n + i)$$

with integer coefficients $a_i \in \mathbb{Z}$. Thus $f_q(n) = \sum_{i=0}^{k'-1} g_i(n) \cdot \theta_i^n$ where $k' \le k$, θ_i are pairwise distinct roots of the polynomial $t^n - \sum_{i=0}^{k-1} a_i \cdot t^i$ and g_i are polynomials of degree not larger than k.

In the subsequent parts we estimate values characterising the exponential growth of the family $\left(|\mathbf{infix}(P_q^*) \cap X^n|\right)_{n \in \mathbb{N}}$. This growth mainly depends on the root of largest modulus among the θ_i and the corresponding polynomial g_i.

First we show that, independently of the quasiperiod q, the root θ_i of largest modulus is always positive and the corresponding polynomial g_i is constant.

In the remainder of this section we use, without explicit reference, known results from the theory of formal power series, in particular about generating functions of languages and codes which can be found in the literature, e.g., in [14,23,24].

5.1. The Subword Complexity of a Regular Star Language

The language P_q^* is a regular star-language of special shape. Here we show that, generally, the number of subwords of regular star-languages grows only exponentially without a polynomial factor. We start with some easily derived relations between the number of words in a regular language and the number of its subwords.

Lemma 6. *If $L \subseteq X^*$ is a regular language then there is an $m \in \mathbb{N}$ such that*

$$|L \cap X^n| \quad \leq \quad |\mathbf{infix}(L) \cap X^n| \quad \leq \quad m \cdot \textstyle\sum_{i=0}^{2m} |L \cap X^{n+i}| \tag{12}$$

If the finite automaton accepting L has m states then for every $w \in \mathbf{infix}(L)$ there are words u, v of length $\leq m$ such that $u \cdot w \cdot v \in L$. Thus as a suitable m one may choose the number of states of an automaton accepting the language $L \subseteq X^*$.

A first consequence of Lemma 6 is that the identity

$$\limsup_{n \to \infty} \sqrt[n]{|L \cap X^n|} = \limsup_{n \to \infty} \sqrt[n]{|\mathbf{infix}(L) \cap X^n|} \tag{13}$$

holds for regular languages $L \subseteq X^*$.

In order to derive the announced exponential growth we use Corollary 4 of [25] which shows that for every regular language $L \subseteq X^*$ there are constants $c_1, c_2 > 0$ and a $\lambda \geq 1$ such that

$$c_1 \cdot \lambda^n \leq |\mathbf{pref}(L^*) \cap X^n| \leq c_2 \cdot \lambda^n\,. \tag{14}$$

A consequence of Lemma 6 is that Equation (14) holds also (with a different constant c_2) for $\mathbf{infix}(L^*)$.

5.2. The Subword Complexity of Q_q

In this part we estimate the value λ_q of Equation (11). In view of Equations (10) and (14) the value λ_q satisfies the inequality $c_1 \cdot \lambda_q^n \leq |\mathbf{infix}(P_q^*) \cap X^n| \leq c_2 \cdot \lambda_q^n$.

As P_q^* is a regular language Equations (11) and (13) show that

$$\lambda_q = \limsup\nolimits_{n \to \infty} \sqrt[n]{|P_q^* \cap X^n|}$$

which is the inverse of the convergence radius $\mathrm{rad}\, \mathfrak{s}_q^*$ of the power series $\mathfrak{s}_q^*(t) := \sum_{n \in \mathbb{N}} |P_q^* \cap X^n| \cdot t^n$. The series $\mathfrak{s}_q^*$ is also known as the structure generating function of the language P_q^*.

Since $\sqrt[*]{P_q}$ is a code, we have $\mathfrak{s}_q^*(t) = \frac{1}{1-\mathfrak{s}_q(t)}$ where $\mathfrak{s}_q(t) := \sum_{v \in \sqrt[*]{P_q}} t^{|v|}$ is the structure generating function of the finite language $\sqrt[*]{P_q}$. As $\mathfrak{s}_q^*$ has non-negative coefficients Pringsheim's theorem shows that $\mathrm{rad}\, \mathfrak{s}_q^* = \lambda_q^{-1}$ is a singular point of $\mathfrak{s}_q^*$. Thus λ_q^{-1} is the smallest root of $1 - \mathfrak{s}_q(t)$. Hence λ_q is the largest positive root of the polynomial $p_q(t) := t^{|q|} - \sum_{v \in \sqrt[*]{P_q}} t^{|q|-|v|}$.

Remark 1. *If the length of $q_0 = \min_{\sqsubseteq} P_q$ does not divide $|q|$ then $p_q(t)$ is the reversed polynomial of $1 - \mathfrak{s}_q(t)$, that is, has as roots exactly the the inverses of the roots of $1 - \mathfrak{s}_q(t)$.*

If $|q_0|$ divides $|q|$ then $q \notin \sqrt[*]{P_q}$ (cf. Corollary 5) and $p_q(t)$ has additionally the root 0 with multiplicity $|q| - |q'|$ where q' is the longest word in $\sqrt[*]{P_q}$.

Summarising our observations we obtain the following.

Lemma 7. *Let $q \in X^* \setminus \{e\}$. Then there are constants $c_{q,1}, c_{q,2} > 0$ such that the structure function of the language* $\mathbf{infix}(P_q^*)$ *satisfies*

$$c_{q,1} \cdot \lambda_q^n \leq |\mathbf{infix}(P_q^*) \cap X^n| \leq c_{q,2} \cdot \lambda_q^n$$

where λ_q is the largest (positive) root of the polynomial $p_q(t)$.

Remark 2. *One could prove Lemma 7 by showing that, for each polynomial $p_q(t)$, its largest (positive) root has multiplicity 1. Referring to Corollary 4 of [25] (see Equation (14)) we avoided these more detailed considerations of a particular class of polynomials.*

Now we are able to formulate our main theorem.

As quasiperiods $q, |q| \leq 2$, have trivially $P_q^* = \{q_0\}^*$, that is, $\lambda_q = 1$, in the sequel we confine our considerations to quasiperiods q of length $|q| \geq 3$, and we will always assume that the first letter of a quasiperiod q is $a \in X$.

Define $Q_{\max} := \{a^n b a^n : n \geq 1\} \cup \{a^n w a^n : |w| = 2, w \neq aa, n \geq 1\}$.

Theorem 3 (Main theorem). *Let $q \in a \cdot X^*, |q| \geq 3, q \notin Q_{\max}$, be a quasiperiod and $n = \lfloor \frac{|q|-1}{2} \rfloor$. Then $\lambda_q < \lambda_{a^n b a^n}$ or $\lambda_q < \lambda_{a^n bb a^n}$ according to whether $|q|$ is odd or even.*

Moreover, $\lambda_w < \lambda_{aba} = \lambda_{aabaa}$ if $w \in a \cdot X^* \setminus \{aba, aabaa\}$.

6. Polynomials

Before proceeding to the proof of our main theorem we derive some properties of polynomials of the form $p(t) = t^n - \sum_{i \in M} t^i$, where $M \subseteq \{i : i \in \mathbb{N} \wedge i < n\}$. This class of polynomials includes the polynomials $p_q(t)$ whose maximal roots λ_q characterise the growth of $\mathbf{infix}(P_q^*)$ as described in Lemma 7. We focus in results which are useful for comparing their maximal roots.

The polynomials $p(t) \in \hat{\mathcal{P}} := \{t^n - \sum_{i \in M} t^i : \emptyset \neq M \subseteq \{0, \ldots, n-1\}\}$ have the following easily verified properties.

$$p(0) \leq 0, p(1) \leq 0, p(2) \geq 1 \text{ and } p(t) < 0 \text{ for } 0 < t < 1. \tag{15}$$

$$\text{If } \varepsilon > 0 \text{ and } p(t') \geq 0 \text{ for some } t' > 0 \text{ then } p((1+\varepsilon) \cdot t') > 0. \tag{16}$$

Since $p(1) \leq 0$ and $p(2) \geq 1$ for $p(t) \in \hat{\mathcal{P}}$, Equation (16) shows that once $p(t') \geq 0$, $t' \geq 1$, the polynomial $p(t)$ has no further root in the interval (t', ∞) and $p(t) \in \hat{\mathcal{P}}$ has exactly one root in the interval $[1, 2)$. This yields the following fundamental property.

Property 1. *If t_0 is the positive root of the polynomial $p(t) \in \hat{\mathcal{P}}$ in $[1,2)$ and $1 \leq t' < 2$ then $p(t') \leq 0$ if and only if $t' \leq t_0$.*

For the roots of maximal modulus we have the following theorem.

Theorem 4 (Cauchy). *Let $p(t) = \sum_{i=0}^{n} a_i \cdot t^i$ be a complex polynomial. Then every root t' of $p(t)$ satisfies $|t'| \leq t_0$ where t_0 is the maximal root of the polynomial $|a_n| \cdot t^n - \sum_{i=0}^{n-1} |a_i| \cdot t^i$.*

This implies the following property of polynomials $p(t) \in \hat{\mathcal{P}}$.

$$\text{If } p(t) = 0 \text{ then } |t| \leq t_0. \tag{17}$$

From Property 1 we derive the following criterion to compare the maximal roots of polynomials in $\hat{\mathcal{P}}$.

Criterion 1. *Let $p_1(t), p_2(t) \in \hat{\mathcal{P}}$ have maximal roots t_1 and t_2, respectively. Then $p_2(t_1) > 0$ if and only if $t_1 > t_2$.*

We conclude this section with a bound on the maximal root of certain polynomials in $\hat{\mathcal{P}}$.

Lemma 8. *Let $p(t) = t^n - \sum_{i=0}^{m} t^i, n > m \geq 1$. Then $p(t) < 0$ for $1 \leq t \leq \sqrt[2n-m]{(m+1)^2}$ and $p(t) > 0$ for $\sqrt[n-m]{m+1} \leq t$.*

Proof. The assertion follows from the inequality $t^n - (m+1) \cdot t^m < p(t) < t^n - (m+1) \cdot t^{m/2}$ when $t > 1$. The part $p(t) < t^n - (m+1) \cdot t^{m/2}$ uses the arithmetic-geometric-means inequality $\sum_{i=0}^{m} t^i > (m+1) \cdot \sqrt[m+1]{\prod_{i=0}^{m} t^i} = (m+1) \cdot t^{m/2}$, and the other part is obvious. □

The following special case is needed below in Lemma 12.

Corollary 10. *If $p(t) = t^n - \sum_{i=0}^{n-3} t^i, n \geq 4$, then $p(t) < 0$ for $1 \leq t \leq \sqrt[n+3]{(n-2)^2}$.*

The subsequent sections are devoted to the proof of our main theorem.

7. Irreducible Quasiperiods

We start with irreducible quasiperiods.

7.1. Extremal Polynomials

The polynomials $p_q(t)$ of irreducible quasiperiods have non-zero coefficients only for $|q|$ and $i < \frac{|q|}{2}$. Therefore we investigate the set

$$\mathcal{P} := \{t^n - \sum_{i \in M} t^i : n \geq 2 \wedge \emptyset \neq M \subseteq \{i : i \leq \tfrac{n-1}{2}\}\}.$$

Let $p_n(t) := t^n - \sum_{i=0}^{\lfloor \frac{n-1}{2} \rfloor} t^i \in \mathcal{P}$.

Property 2. *Let $p(t) \in \mathcal{P}$ a polynomial of degree $n \geq 3$. Then $p_n(t) \leq p(t)$ for $t \in [1,2]$, and $p_n(t)$ has the largest positive root among all polynomials of degree n in $\mathcal{P}$.*

Proof. This follows from $t^n - \sum_{i=0}^{\lfloor \frac{n-1}{2} \rfloor} t^i < p(t)$ for $p(t) \in \mathcal{P} \setminus \{p_n(t) : n \geq 3\}$ when $1 < t \leq 2$ and Criterion 1. □

Observe that, for $n \geq 1$,

$$p_{2n+1}(t) = t^{2n+1} - \sum_{i=0}^{n} t^i \text{ and } p_{2n+2}(t) = t^{2n+2} - \sum_{i=0}^{n} t^i.$$

Moreover, the words $a^n b a^n \in Q_{\max}$ and $a^n w a^n \in Q_{\max}, w \in \{xb, bx\}, x \in X$ are the quasiperiods corresponding to the extremal polynomials $p_{2n+1}(t) \in \mathcal{P}$ and $p_{2n+2}(t) \in \mathcal{P}$, respectively.

Lemma 9. $Q_{\max} := \{q : q \in a \cdot X^* \wedge |q| \geq 3 \wedge p_q(t) = p_{|q|}(t)\}$

Proof. If $q \in Q_{\max}$ then obviously $p_q(t) = p_{|q|}(t)$. Conversely, if $p_q(t) = t^{|q|} - \sum_{v \in \sqrt[*]{P_q}} t^{|q|-|v|} = p_{|q|}(t)$ then $\sqrt[*]{P_q} = \{v : v \sqsubseteq q \wedge |v| > \frac{|q|}{2}\}$. Then, in view of $q \sqsubseteq v \cdot q$, every prefix $w \sqsubseteq q$ of length $|w| < \frac{|q|}{2}$ is also a suffix of q. This is possible only for $q \in Q_{\max}$ or $q \in \{a\}^*$. □

In the sequel the positive root of $p_n(t)$ is denoted by λ_n. From Criterion 1 we obtain immediately.

Property 3. *Let $t \geq 1$. We have $t < \lambda_n$ if and only if $p_n(t) < 0$.*

Then Property 2 implies the following.

Theorem 5. *If $q \in a \cdot X^*$, $|q| \geq 3$, is an irreducible quasiperiod then $\lambda_q \leq \lambda_{|q|}$, and $\lambda_q = \lambda_{|q|}$ if and only if $q \in Q_{\max}$.*

7.2. The Ordering of the Maximal Roots λ_n

Before we proceed to the case of reducible quasiperiods we determine the ordering of the maximal roots λ_n. This will not only be interesting for itself but also useful for proving $\lambda_q < \lambda_{|q|}$ when q is reducible (see Equation (28) below).

The extremal polynomials $p_n(t), n \geq 2$, satisfy the following general relations (By convention, $\sum_{i=k}^{m} a_i = 0$ if $k > m$).

$$
\begin{aligned}
t \cdot p_{2n}(t) - 1 &= p_{2n+1}(t)\,, && (18)\\
p_{2n+2}(t) - t^2 \cdot p_{2n}(t) &= t^{n+1} - t - 1\,, && (19)\\
t^{n-2} \cdot p_{2n+1}(t) - (t^n + 1) \cdot p_{2n-1}(t) &= \textstyle\sum_{i=0}^{n-3} t^i, \text{ and} && (20)\\
t^{n-2} \cdot p_{2n+3}(t) - (t^{n+1} + 1) \cdot p_{2n}(t) &= -t^n + \textstyle\sum_{i=0}^{n-3} t^i\,. && (21)
\end{aligned}
$$

Lemma 10. *The polynomials $t^3 - t - 1$ and $t^5 - t^2 - t - 1 = (t^2+1) \cdot (t^3 - t - 1)$ have largest positive roots $\lambda_3 = \lambda_5$ among all polynomials in $\mathcal{P}$, $\lambda_5 > \lambda_4$ and $\lambda_{2n-1} > \lambda_{2n+1} > \lambda_{2n}$ for $n \geq 3$.*

Proof. From Equation (18) we have $p_{2n+1}(\lambda_{2n}) = -1 < 0$ and, therefore, $\lambda_{2n} < \lambda_{2n+1}$ when $n \geq 1$.

Similarly, Equation (20) yields $p_{2n+1}(\lambda_{2n-1}) = \lambda_{2n-1}^{-(n-2)} \cdot \sum_{i=0}^{n-3} \lambda_{2n-1}^i > 0$ which implies $\lambda_{2n+1} < \lambda_{2n-1}$ for $n \geq 3$ and $\lambda_3 = \lambda_5$ when $n = 2$. □

The largest (positive) root λ_3 of the polynomial $t^3 - t - 1$ is also known as the smallest Pisot-Vijayaraghavan number.

So far we have ordered the 'odd' roots: $\lambda_3 = \lambda_5 > \lambda_7 > \lambda_9 > \cdots$. Next we are going to investigate the ordering of the 'even' roots λ_{2n}, $n \geq 2$.

To this end we derive the following bounds.

Lemma 11.

1. $\sqrt[3n+1]{n^2} \leq \lambda_{2n} \leq \sqrt[n+1]{n}$ *and* $\sqrt[3n-1]{n^2} \leq \lambda_{2n-1} \leq \sqrt[n]{n}$ *for* $n \geq 2$.
2. *Let* $n \geq 5$. *Then* $\lambda_{2n} \geq \sqrt[n-1]{2}$.

Proof. 1. follows from Lemma 8.

2. We calculate $p_{2n}(\sqrt[n-1]{2}) = 4 \cdot \sqrt[n-1]{4} - \sum_{i=0}^{n-1} \sqrt[n-1]{2^i} \leq 4 \cdot \sqrt[4]{4} - (2 + (n-1)) = 4 \cdot \sqrt{2} - (n+1) < 0$ if $n \geq 5$ and the assertion follows with Property 1. □

Remark 3. The lower bound of Lemma 11.2 does not exceed the lower bound in Lemma 11.1. However, the latter is more convenient for the purposes of Lemma 12.

Lemma 12. *If $n \geq 5$ then $\lambda_{2n-2} > \lambda_{2n}$ and $\lambda_{2n} > \lambda_{2n+3}$.*

Proof. If $t \geq \sqrt[n-1]{2}$ then $t^n - t - 1 \geq t - 1 > 0$. Consequently, Equation (19) and Lemma 11.2 imply $p_{2n-2}(\lambda_{2n}) < 0$ whence $\lambda_{2n} < \lambda_{2n-2}$.

If $n \geq 5$ we have $\sqrt[n+1]{n} \leq \sqrt[n+3]{(n-2)^2}$ and, following Lemma 11.1 $\lambda_{2n} \leq \sqrt[n+3]{(n-2)^2}$. Then Equation (21) yields $-\lambda_{2n} \cdot p_{2n+3}(\lambda_{2n}) = \lambda_{2n}^n - \sum_{i=0}^{n-3} \lambda_{2n}^i$, and Corollary 10 shows $p_{2n+3}(\lambda_{2n}) > 0$ whence $\lambda_{2n} > \lambda_{2n+3}$. □

Since $p_8(\sqrt[3]{2}) > 0$, the proof of Lemma 12 cannot be applied to lower values of n. Thus it remains to establish the order of the λ_i for $i \leq 13$. To this end, we consider some special identities and use Criterion 3 and Lemma 12.

$$p_{12}(t) - (t^8 + t^5 + t^4 + t^2 + t) \cdot p_4(t) = t^2 - 1, \text{ and} \tag{22}$$
$$p_{13}(t) - t \cdot (t^8 + t^5 + t^4 + t^2 + t) \cdot p_4(t) = t^3 - t - 1 = p_3(t). \tag{23}$$

Lemma 13. $\lambda_8 > \lambda_{10} > \lambda_{13} > \lambda_4 > \lambda_{12}$

Proof. Lemma 12 shows $\lambda_8 > \lambda_{10} > \lambda_{13}$. Equation (22) yields $p_{12}(\lambda_4) = \lambda_4^2 - 1 > 0$ whence $\lambda_4 > \lambda_{12}$, and Equation (23) yields $p_{13}(\lambda_4) = p_3(\lambda_4) < 0$, that is, $\lambda_{13} > \lambda_4$. This shows our assertion. □

For the remaining part we consider the identities

$$t^2 \cdot p_{11}(t) - (t^5 + 1) \cdot p_8(t) = -t^4 + t + 1 = -p_4(t), \tag{24}$$
$$p_{11}(t) - (t^5 + 1) \cdot p_6(t) = t^3 \cdot p_4(t), \text{ and} \tag{25}$$
$$t \cdot p_9(t) - (t^4 + 1) \cdot p_6(t) = -t^3 + 1. \tag{26}$$

Lemma 14. $\lambda_9 > \lambda_6 > \lambda_{11} > \lambda_8$

Proof. We use Equations (24)–(26). Then $p_{11}(\lambda_8) = -p_4(\lambda_8) < 0$ implies $\lambda_{11} > \lambda_8$, $p_{11}(\lambda_6) = \lambda_6^3 \cdot p_4(\lambda_6) > 0$ implies $\lambda_6 > \lambda_{11}$, and, finally, $\lambda_6 \cdot p_9(\lambda_6) = -\lambda_6^3 + 1 < 0$ implies $\lambda_9 > \lambda_6$. □

Now Lemma 10, 12–14 yield the complete ordering of the values λ_n.

Theorem 6. *Let $\lambda_n, n \geq 3$, be the maximal root of the polynomial $p_n(t)$. Then the overall ordering of the values λ_n starts with*

$$\lambda_3 = \lambda_5 > \lambda_7 > \lambda_9 > \lambda_6 > \lambda_{11} > \lambda_8 > \lambda_{10} > \lambda_{13} > \lambda_4 > \lambda_{12}$$

and continues as follows $\lambda_{2n+1} > \lambda_{2n} > \lambda_{2n+3}, n \geq 7$.

In connection with Proposition 6 and Corollary 7 we obtain that the Pisot-Vijayaraghavan number $\lambda_3 = \lambda_5$ is an overall upper bound on the values λ_q.

Corollary 11. *If $q \in X^*, |q| \geq 3$, then $\lambda_q \leq \lambda_3 = \lambda_5$.*

From Lemma 11.1 we obtain immediately.

Corollary 12. *Let $M \subseteq \mathbb{N} \setminus \{0, 1, 2\}$ be infinite. Then* $\inf\{\lambda_i : i \in M\} = 1$.

8. Reducible Quasiperiods

Reducible quasiperiods q have a repeated prefix $q_0 = \min_{\sqsubseteq} P_q$ with $|q_0| \leq |q|/2$ and a repetition factor $k \geq 2$ such that $q = q_0^k \cdot \bar{q}$ where $\bar{q} \sqsubset q_0$. Moreover $|\bar{q}| < |q_0| \leq |q|/2$. Observe that q_0 is primitive.

We shall consider three cases depending on the relation between the lengths $n = |q|$, $\ell = |q_0|$, the length of the suffix $|\bar{q}| < |q_0|$ and the repetition factor $k \geq 2$.

IN the first case $|q_0| + |\bar{q}| \leq 2$, in view of $\bar{q} \sqsubset q_0$, we have necessarily $\bar{q} = e$ and $q \in a^* \cup \{ab\}^*$, $a, b \in X, a \neq b$ and, therefore, $Q_q = \{q_0\}^*$ and $\lambda_q = 1$.

Let now $|q_0| + |\bar{q}| \geq 3$. We divide the remaining cases according to the additional requirement $|q| - 2|q_0| \geq 3$ and its complementary one $|q| - 2|q_0| \leq 2$. In the latter case we have necessarily $k = 2$ and $|\bar{q}| \leq 2$.

8.1. The Case $|q_0| + |\bar{q}| \geq 3 \wedge |q| - 2|q_0| \geq 3$

Thus, the preceding consideration shows that we have $|\bar{q}| \geq 3$ (in particular, if $q = q_0^2 \cdot \bar{q}$) or the repetition factor $k \geq 3$. This implies $|q| = 7$ (where $q = (ab)^3 a$) or $|q| \geq 9$. From Equation (6) we have

$$\sqrt[*]{P_q} \subseteq \{q_0\} \cup \{v : v \sqsubseteq q \wedge |v| > |q| - |q_0| + 1\} \tag{27}$$

This implies that for $|q_0| \leq |q|/2$ the polynomials $p_q(t)$ have non-zero coefficients only for $|q| = n$, $|q| - |q_0| = n - \ell$ and $i < |q_0| - 1$, that is, are of the form $p_q(t) = t^n - t^{n-\ell} - \sum_{i \in M_q} t^i$ where $M_q \subseteq \{i : i < \ell - 1\}$. Therefore, in the sequel we consider the positive roots of polynomials in

$$\mathcal{P}_{\text{red}} := \{t^n - t^{n-\ell} - \sum_{i \in M} t^i : n \geq 1 \wedge \ell \leq \frac{n}{2} \wedge M \subseteq \{i : i < \ell - 1\}\}$$

Let $p_{n,\ell}(t) := t^n - t^{n-\ell} - \sum_{i=0}^{\ell-2} t^i \in \mathcal{P}_{\text{red}}$ and $\lambda_{n,\ell}$ be its maximal root. (In the preceding paper [8] we used a slightly different definition of $\mathcal{P}_{\text{red}}$, and, therefore, of $p_{n,\ell}(t)$ and $\lambda_{n,\ell}$.) Similar to Property 2, Criterion 3 and Theorem 5 we have the following.

Property 4. *Let $n \geq 3, \ell \leq \frac{n}{2}$ and $p(t) \in \mathcal{P}_{\text{red}}$. Then $p(t) \geq p_{n,\ell}(t)$ for $t \in [1,2]$, and $p_{n,\ell}(t)$ has the largest positive root among all polynomials of degree n and parameter ℓ in $\mathcal{P}_{\text{red}}$.*

Lemma 15. *If q, $|q| = n$, is a quasiperiod with $|q_0| = \ell \leq n/2$ then $p_q(t) \geq p_{n,\ell}(t)$ for $t \geq 1$, in particular, $\lambda_q \leq \lambda_{n,\ell}$.*

Remark 4. *In contrast to Property 2 not for every polynomial $p_{n,\ell}(t)$ there is a quasiperiod q such that $p_{n,\ell}(t) = p_q(t)$, see Remark 5 below.*

We have the following relation between the polynomials $p_n(t)$ and $p_{n,\ell}(t)$.

$$p_n(t) - t^\ell \cdot p_{n-2\ell}(t) = p_{n,\ell}(t) - t^{\ell-1}, \text{ for } n - 2\ell \geq 3 \tag{28}$$

This yields

Corollary 13. *Let $n - 2 \cdot \ell \geq 3$. If $\lambda_n < \lambda_{n-2\ell}$ then $\lambda_{n,\ell} < \lambda_n$.*

Proof. If $\lambda_n < \lambda_{n-2\ell}$ then $p_{n-2\ell}(\lambda_n) < p_{n-2\ell}(\lambda_{n-2\ell}) = 0$. Thus $p_{n,\ell}(\lambda_n) = -\lambda_n^\ell \cdot p_{n-2\ell}(\lambda_n) + \lambda_n^{\ell-1} > 0$, that is, $\lambda_n > \lambda_{n,\ell}$. □

Next we show the relation $\lambda_q < \lambda_{|q|}$ for all quasiperiods q having $|q_0| \leq |q|/2$ and $|q_0| + |\bar{q}| \geq 3$.

Lemma 16. *Let $|q| - 2|q_0| \geq 3$ and $|q_0| + |\bar{q}| \geq 3$. Then $\lambda_q < \lambda_{|q|}$.*

Proof. Above we have shown that $|q| - 2|q_0| \geq 3$ and $|q_0| + |\bar{q}| \geq 3$ imply $|q| \geq 7$ or $|q| \geq 10$ according to whether $|q|$ is odd or even.

The ordering of Theorem 6 and Corollary 13 show $\lambda_n > \lambda_{n,\ell}$ for all odd values $n \geq 7$ and for all even values $n \geq 12$.

It remains to consider the exceptional case when $n = |q| = 10$. Here $|q| - 2|q_0| \geq 3$ and $|q_0| + |\bar{q}| \geq 3$ imply $\ell = |q_0| = 3$. Consider $p_{10,3}(t) = t^{10} - t^7 - t - 1 = p_{10}(t) - t^2 \cdot p_5(t)$. From $\lambda_5 > \lambda_{10}$ and $p_{10}(\lambda_{10}) = 0$ we have $p_{10,3}(\lambda_{10}) = -\lambda_{10}^2 \cdot p_5(\lambda_{10}) > 0$, that is, $\lambda_{10,3} < \lambda_{10}$. □

Remark 5. *Equation (6) shows that for* $n = |q| = 10$ *and* $\ell = |q_0| = 3$ *we have* $\sqrt[*]{P_q} = \{q_0, q\}$, *that is,* $p_q(t) = t^{10} - t^7 - 1$. *Thus there is no quasiperiod* q *such that* $p_q(t) = p_{10,3}(t) = t^{10} - t^7 - t - 1$.

8.2. *The Case* $|q_0| + |\bar{q}| \geq 3 \wedge |q| - 2|q_0| \leq 2$

This amounts to $|q| = 2 \cdot |q_0| + |\bar{q}|$ where $|\bar{q}| \in \{0, 1, 2\}$.

Here we have to go into more detail and to take into consideration also the reduced quasiperiod $\hat{q} = q_0 \cdot \bar{q}$ of q and its repeated prefix $\hat{q}_0 = \min_{\sqsubseteq} P_{\hat{q}}$. Observe that both repeated prefixes $q_0, \hat{q}_0$ are primitive.

For $q = q_0^k \cdot \bar{q}, k \geq 2$, we have from Equations (7) and (8)

$$p_q(t) \in \left\{ t^{|q|} - t^{|q| - |q_0|} - \textstyle\sum_{i \in M} t^i : M \subseteq \{0, \ldots, |\hat{q}| - |\hat{q}_0|\} \right\}.$$

Observe that $|\hat{q}_0| > |\bar{q}|$ (in view of Lemma 3 even $|\hat{q}_0| > |\bar{q}| + 1$ if $\hat{q}_0 \neq q_0$) and thus $|\hat{q}| - |\hat{q}_0| = |q_0| - (|\hat{q}_0| - |\bar{q}|) < |q_0|$.

Let $\mathcal{P}'_{\mathrm{red}} := \left\{ t^n - t^\ell - \sum_{i \in M} t^i : n > \ell > j \wedge M \subseteq \{0, \ldots, \ell - j\} \right\}$ and $p_{n,\ell,j}(t) = t^n - t^\ell - \sum_{i=0}^{\ell-j} t^i$. Here the parameter j corresponds to the value $|\hat{q}_0| - |\bar{q}|$. Then similar to Property 4 and Lemma 15 we have

Property 5. *Let* $n, \ell \geq 3, \ell \leq \frac{n}{2}, \ell > j$, *and* $p(t) \in \mathcal{P}'_{\mathrm{red}}$. *Then* $p(t) \geq p_{n,\ell,j}(t)$ *for* $t \in [1, 2]$, *and* $p_{n,\ell,j}(t)$ *has the largest positive root among all polynomials of degree n and parameters* ℓ *and* j *in* $\mathcal{P}'_{\mathrm{red}}$.

Lemma 17. *If* $q, |q| = n$, *is a quasiperiod with* $|q_0| = \ell \leq n/2$ *and* $|\hat{q}_0| - |\bar{q}| \geq j$ *then* $p_q(t) \geq p_{n,\ell,j}(t)$ *for* $t \geq 1$, *in particular,* $\lambda_q \leq \lambda_{n,\ell,j}$.

We consider the cases $|\bar{q}| \in \{0, 1, 2\}$ separately. In the sequel we shall make use of the relation

$$t^3 - t^2 - 1 \leq t^2 - t - 1 < 0 \text{ for } 1 \leq t \leq \lambda_3 = \max\{\lambda_n : n \in \mathbb{N}\}. \quad (29)$$

8.2.1. The Case $q = q_0^2 \wedge |\bar{q}| = 0$

As shown above the case $|q_0| \leq 2$ and $|\bar{q}| = 0$ amounts to $\lambda_q = 1$. Thus we may consider only the case when $|q_0| \geq 3$. Here we have the following relation between $p_{2\ell}(t)$ and $p_{2\ell,\ell,3}(t)$.

$$p_{2\ell}(t) - p_{2\ell,\ell,3}(t) = t^{\ell-2}(t^2 - t - 1) \quad (30)$$

Lemma 18. *If* $q = q_0^2$ *and* $|q_0| = \ell \geq 3$ *then* $\lambda_q < \lambda_{|q|}$.

Proof. First we suppose $|\hat{q}_0| \geq 3$. Then $|\hat{q}_0| - |\bar{q}| \geq 3$, and Property 5 and Lemma 17 yield $p_q(t) \geq p_{2\ell,\ell,3}(t)$ for $t \in [1, 2]$. Now Equations (29) and (30) show $p_q(\lambda_{2\ell}) \geq p_{2\ell,\ell,3}(\lambda_{2\ell}) = -\lambda_{2\ell}^{\ell-2}(\lambda_{2\ell}^2 - \lambda_{2\ell} - 1) > 0$, that is $\lambda_q < \lambda_{2\ell}$.

It remains to consider $1 \leq |\hat{q}_0| \leq 2$. If $\hat{q}_0 \in a^*$ then $q_0 = a^\ell$ which is not primitive. Thus $\hat{q}_0 = ab$ and, since q_0 is primitive, $q_0 = (ab)^m a$, $m \geq 1$ whence $q = q_0^2 = (ab)^m a \cdot (ab)^m a$.

We obtain $\sqrt[*]{P_q} = \{(ab)^m a \cdot (ab)^i : i = 0, \ldots, m\}$ and, consequently, $p_q(t) = t^{4m+2} + \sum_{i=0}^m t^{2i+1}$. Then (Observe again $\sum_{i=k}^m a_i = 0$ if $k > m$).

$$\begin{aligned} p_q(t) - p_{4m+2}(t) &= -t^{2m+1} + \sum_{i=0}^{m} t^{2i} = -t^{2m+1} + t^{2m} + t^{2m-2} + \sum_{i=0}^{m-2} t^{2i} \\ &= -t^{2m-2} \cdot (t^3 - t^2 - 1) + \sum_{i=0}^{m-2} t^{2i}, \end{aligned}$$

and from Equation (29) we obtain $p_q(\lambda_{4m+2}) \geq -\lambda_{4m+2}^{2m-2}(\lambda_{4m+2}^3 - \lambda_{4m+2}^2 - 1) > 0$. □

8.2.2. The Case $q = q_0^2 \cdot \bar{q} \wedge |\bar{q}| = 1$

Here we have the following relation between $p_{2\ell+1}(t)$ and $p_{2\ell+1,\ell,2}(t)$.

$$p_{2\ell+1}(t) - p_{2\ell+1,\ell,2}(t) = t^{\ell-1}(t^2 - t - 1) \tag{31}$$

Lemma 19. *If* $q = q_0^2 \cdot a, a \in X$, *then* $\lambda_q < \lambda_{|q|}$.

Proof. First we suppose $|\hat{q}_0| - |\bar{q}| \geq 2$. Then $\ell = |q_0| \geq |\hat{q}_0| \geq 3$, and Property 5 and Equation (31) yield $p_q(\lambda_{2\ell+1}) \geq p_{2\ell+1,\ell,2}(\lambda_{2\ell+1}) = p_{2\ell+1}(\lambda_{2\ell+1}) - \lambda_{2\ell+1}^{\ell-1}(\lambda_{2\ell+1}^2 - \lambda_{2\ell+1} - 1)$. The assertion $p_q(\lambda_{2\ell+1}) > 0$, that is $\lambda_q < \lambda_{2\ell+1}$ follows from Equation (29).

It remains to consider $|\hat{q}_0| = 2$. By Lemma 3 $\hat{q}_0 = q_0$ implies $|\hat{q}_0| > |\bar{q}| + 1 = 2$. Hence $\hat{q}_0 = q_0 = ab$, $q = ababa$ and $p_q(t) = t^5 - t^3 - 1 = t^2 \cdot p_3(t) + t^2 - 1$. Then $\lambda_{ababa} < \lambda_5$ follows from $\lambda_5 = \lambda_3$ and $p_q(\lambda_5) = \lambda_5^2 - 1 > 0$. □

8.2.3. The Case $q = q_0^2 \cdot \bar{q} \wedge |\bar{q}| = 2$

Here we have the following relation between $p_{2\ell+2}(t)$ and $p_{2\ell+2,\ell,2}(t)$.

$$p_{2\ell+2}(t) - p_{2\ell+2,\ell,2}(t) = t^{\ell-1}(t^3 - t - 1) = t^{\ell-1} \cdot p_3(t) \tag{32}$$

Lemma 20. *If* $q = q_0^2 \cdot \bar{q}$ *with* $|\bar{q}| = 2$ *then* $\lambda_q < \lambda_{|q|}$.

Proof. First we suppose $|\hat{q}_0| \geq 4$. Then Property 5, Equation (32) and $\lambda_{2\ell+2} < \lambda_3$ yield $p_q(\lambda_{2\ell+2}) \geq p_{2\ell+2,\ell,2}(\lambda_{2\ell+2}) = -\lambda_{2\ell+2}^{\ell-1} \cdot p_3(\lambda_{2\ell+2}) > 0$, that is, $\lambda_q < \lambda_{2\ell+2}$.

It remains to consider $|\hat{q}_0| = 3$. If $\hat{q}_0 \neq q_0$ Lemma 3 implies $|\hat{q}_0| > |\bar{q}| + 1$. Consequently, $\hat{q}_0 = q_0$. Then $|q_0| = 3$ and $|q| = 8$, and Equation (6) yields $\sqrt[*]{P_q} \subseteq \{q_0, v, q\}$ where $v \sqsubset q$ and $|v| = |q| - 1 = 7$. Thus $p_q(t) \geq t^8 - t^5 - t - 1 = p_8(t) - t^2 \cdot p_3(t)$ for $1 \leq t \leq \lambda_3$.

This shows $p_q(\lambda_8) \geq -\lambda_8^2 \cdot p_3(\lambda_8) > 0$, that is, $\lambda_q < \lambda_8$. □

Summarising, the results of Section 8 yield

Theorem 7. *If* $q \in X^*$, $|q| \geq 3$, *is a reducible quasiperiod then* $\lambda_q < \lambda_{|q|}$.

Our main theorem (Theorem 3) then follows from Theorems 5 and 7.

Together with Corollary 12 our theorem yields a new proof of a theorem of [5] which shows that multi-scale quasiperiodic infinite words have zero topological entropy. In [5] a *multi-scale quasiperiodic infinite word* is a quasiperiodic infinite word which admits infinitely many quasiperiods.

9. Concluding Remark

In this paper we dealt with the function $f(\xi, n) = |\mathbf{infix}(\xi) \cap X^n|$ for quasiperiodic ω-words. Their factor complexity (or topological entropy) is defined as $\tau(\xi) := \lim_{n\to\infty} \frac{\log_{|X|} |\mathbf{infix}(\xi) \cap X^n|}{n}$ (e.g., [4], Section 4.2.2 or [5,22]). Thus the upper bound for $\xi \in P_q^\omega$ is $\log_{|X|} \lambda_q \leq \log_{|X|} t_P$ which is bounded away from the value 1 for almost periodic ω-words.

Along with the subword complexity in [5] the Kolmogorov complexity of quasiperiodic ω-words was addressed. Obviously, subword complexity upper bounds Kolmogorov complexity (e.g., [22]). Since the ω-languages P_q^ω are regular ones, the results of [22] show that there are ω-words $\xi \in P_q^\omega$ whose Kolmogorov complexity achieves their subword complexity. Moreover, as $P_q^\omega = q \cdot R_q^\omega$ where R_q^ω is a finite prefix code, the results of [22,26,27] give more detailed bounds for most complex quasiperiodic ω-words w.r.t. several notions of Kolmogorov complexity [28].

Funding: This research received no external funding.

Institutional Review Board Statement: Not applicable.

Informed Consent Statement: Not applicable.

Data Availability Statement: Not applicable.

Conflicts of Interest: The author declares no conflict of interest.

References

1. Marcus, S. Bridging two hierarchies of infinite words. *J. UCS* **2002**, *8*, 292–296.
2. Marcus, S. Quasiperiodic infinite words (column: Formal language theory). *Bull. EATCS* **2004**, *82*, 170–174.
3. Marcus, S.; Păun, G. Infinite (almost periodic) words, formal languages and dynamical systems. *Bull. EATCS* **1994**, *54*, 224–231.
4. Cassaigne, J.; Nicolas, F. Factor complexity. Combinatorics, automata and number theory. In *Encyclopedia of Mathematics and Its Applications*; Cambridge University Press: Cambridge, UK, 2010; Volume 135, pp. 163–247.
5. Monteil, T.; Marcus, S. Quasiperiodic infinite words: Multi-scale case and dynamical properties. *Mathematics* **2006**, arXiv:math/0603354.
6. Polley, R.; Staiger, L. The maximal subword complexity of quasiperiodic infinite words. In Proceedings of the Twelfth Annual Workshop on Descriptional Complexity of Formal Systems, Saskatoon, SK, Canada, 8–10 August 2010.
7. Staiger, L. Quasiperiods of infinite words. In *Mathematics almost Everywhere. In Memory of Solomon Marcus*; Bellow, A., Calude, C.S., Zamfirescu, T., Eds.; World Scientific: Hackensack, NJ, USA, 2018; pp. 17–36.
8. Staiger, L. On the generative power of quasiperiods. In Lecture Notes in Computer Science, Proceedings of the 22nd International Conference Descriptional Complexity of Formal Systems, Vienna, Austria, 24–26 August 2020; Jirásková, G., Pighizzini, G., Eds.; Springer: Cham, Switzerland, 2020; Volume 12442, pp. 219–230.
9. Levé, F.; Richomme, G. Quasiperiodic infinite words: Some answers (column: Formal language theory). *Bull. EATCS* **2004**, *84*, 128–138.
10. Apostolico, A.; Ehrenfeucht, A. Efficient detection of quasiperiodicities in strings. *Theor. Comput. Sci.* **1993**, *119*, 247–265. [CrossRef]
11. Apostolico, A.; Farach, M.; Iliopoulos, C.S. Optimal superprimitivity testing for strings. *Inform. Process. Lett.* **1991**, *39*, 17–20. [CrossRef]
12. Mouchard, L. Normal forms of quasiperiodic strings. *Theoret. Comput. Sci.* **2000**, *249*, 313–324. [CrossRef]
13. Brzozowski, J.A. Roots of star events. *J. ACM* **1967**, *14*, 466–477. [CrossRef]
14. Berstel, J.; Perrin, D. Theory of codes. In *Pure and Applied Mathematics*; Academic Press Inc.: Orlando, FL, USA, 1985; Volume 117.
15. Shyr, H.-J. *Free Monoids and Languages*, 3rd ed.; Hon Min Book Company: Taichung, Taiwan, 2001.
16. Lothaire, M. *Combinatorics on Words*, 2nd ed.; Cambridge University Press: Cambridge, UK, 1997; Volume 17.
17. Fine, N.J.; Wilf, H.S. Uniqueness theorems for periodic functions. *Proc. Am. Math. Soc.* **1965**, *16*, 109–114. [CrossRef]
18. Bruyère, V.; Wang, L.M.; Zhang, L. On completion of codes with finite deciphering delay. *Eur. J. Combin.* **1990**, *11*, 513–521. [CrossRef]
19. Fernau, H.; Reinhardt, K.; Staiger, L. Decidability of code properties. *Theor. Inform. Appl.* **2007**, *41*, 243–259. [CrossRef]
20. Staiger, L. On infinitary finite length codes. *RAIRO Inform. Théor. Appl.* **1986**, *20*, 483–494. [CrossRef]
21. Devolder, J.; Latteux, M.; Litovsky, I.; Staiger, L. Codes and infinite words. *Acta Cybernet.* **1994**, *11*, 241–256.
22. Staiger, L. Kolmogorov complexity and Hausdorff dimension. *Inform. Comput.* **1993**, *103*, 159–194. [CrossRef]
23. Berstel, J.; Reutenauer, C. Rational series and their languages. In *EATCS Monographs on Theoretical Computer Science*; Springer: Berlin/Heidelberg, Germany, 1988; Volume 12.
24. Salomaa, A.; Soittol, M. *Automata-Theoretic Aspects of Formal Power Series*; Springer: New York, NY, USA, 1978.
25. Staiger, L. The entropy of finite-state ω-languages. *Probl. Control. Inform. Theory/Probl. Upravlen. Teor. Inform.* **1985**, *14*, 383–392.
26. Mielke, J.; Staiger, L. On oscillation-free ε-random sequences II. In *Computability and Complexity in Analysis*; Dagstuhl Seminar Proceedings; Bauer, A., Hertling, P., Ko, K.-I., Eds.; Schloss Dagstuhl-Leibniz-Zentrum für Informatik: Wadern, Germany, 2009; Volume 09003.
27. Staiger, L. Bounds on the Kolmogorov complexity function for infinite words. In *Information and Complexity*; World Scientific Series in Information Studies; Chapter 8; Burgin, M., Calude, C.S., Eds.; World Scientific: Hackensack, NJ, USA, 2017; Volume 6, pp. 200–224.
28. Uspensky, V.A.; Shen, A. Relations between varieties of Kolmogorov complexities. *Math. Syst. Theory* **1996**, *29*, 271–292. [CrossRef]

axioms

MDPI

Article

P Systems with Evolutional Communication and Division Rules

David Orellana-Martín [1,*], Luis Valencia-Cabrera [1,2] and Mario J. Pérez-Jiménez [1,2]

1 Research Group on Natural Computing, Department of Computer Science and Artificial Intelligence, Universidad de Sevilla, Avda. Reina Mercedes s/n, 41012 Sevilla, Spain; lvalencia@us.es (L.V.-C.); marper@us.es (M.J.P.-J.)
2 Smart Computer Systems Research and Engineering Lab (SCORE), Research Institute of Computer Engineering (I3US), Universidad de Sevilla, 41012 Sevilla, Spain
* Correspondence: dorellana@us.es

Abstract: A widely studied field in the framework of membrane computing is computational complexity theory. While some types of P systems are only capable of efficiently solving problems from the class **P**, adding one or more syntactic or semantic ingredients to these membrane systems can give them the ability to efficiently solve presumably intractable problems. These ingredients are called to form a frontier of efficiency, in the sense that passing from the first type of P systems to the second type leads to passing from non-efficiency to the presumed efficiency. In this work, a solution to the SAT problem, a well-known **NP**-complete problem, is obtained by means of a family of recognizer P systems with evolutional symport/antiport rules of length at most (2,1) and division rules where the environment plays a passive role; that is, P systems from $\widehat{\mathcal{CDEC}}(2,1)$. This result is comparable to the one obtained in the tissue-like counterpart, and gives a glance of a parallelism and the non-evolutionary membrane systems with symport/antiport rules.

Keywords: membrane computing; computational complexity theory; **P** vs. **NP** problem; evolutional communication; symport/antiport

MSC: 68Q07; 68Q15

Citation: Orellana-Martín, D.; Valencia-Cabrera, L.; Pérez-Jiménez, M.J. P Systems with Evolutional Communication and Division Rules. *Axioms* **2021**, *10*, 327. https://doi.org/10.3390/axioms10040327

Academic Editors: Cristian S. Calude and Gheorghe Păun

Received: 1 October 2021
Accepted: 24 November 2021
Published: 30 November 2021

1. Introduction

Membrane computing is a bio-inspired paradigm of computation, based on the structure and behavior of living cells. Introduced in 1998 by Gh. Păun [1], giving birth to devices known as *membrane systems* or *P systems*. There are several different types of P systems, but three of them are specially studied: cell-like membrane systems [1], whose tree-like structure characterizes the relation between its regions; tissue-like membrane systems [2], defined as a set of cells that can interact between them and with the environment, and neural-like membrane systems [3], having an explicitly defined directed graph as a relation of the neurons through synapses. The last paradigm is being intensely studied in practical applications, and different variants have been created to their use in different fields [4–6] The paradigm of membrane computing is very wide, covering topics from theory [7,8] to applications [9–11], dedicating a branch to simulators and in silico implementations [12].

A widely studied question in this framework from the very beginning is which kind of problems can be solved by means of membrane systems. Membrane systems can differ in the type of objects with which they can compute (e.g., symbols, strings, matrices), the type of relation between the regions (e.g., hierarchical structure, directly connected membranes, cells implicitly connected by the rules) and the rules governing the computation of the system (e.g., object evolution rules, symport/antiport rules, division rules), among others. This variety of ingredients can change not only the type of problem that a P system can solve, but how efficiently can it solve a certain problem. More precisely, decision problems are usually studied in the field of computational complexity theory in order to classify them in complexity classes that contain problems that can be solved with a similar amount of computational resources [13].

Recognizer membrane systems [14] are P systems with certain ingredients, such as two special objects `yes` and `no`, and requisites, such as all the computations halt and return the same result. The design of a family of recognizer membrane systems of a certain type $\mathcal{R}$ solving certain decision problems can reveal which kind of problems can be solved efficiently by means of that class of P systems.

A widely studied type of membrane systems in the field of computational complexity theory is the framework of tissue P systems with symport/antiport rules. In [15], a polynomial-time solution to SAT was designed by means of a family of recognizer tissue P systems with symport/antiport rules of length at most five and division rules. In this type of P system, the length is defined as the number of objects implied in the symport/antiport rules of the system (e.g., the length of the rule $(i, u/v, j)$ is $|u| + |v|$). This result was eventually improved in [16], where the maximum number of objects implied in a communication rule was two. If only one object was allowed in communication rules, then only tractable problems could be efficiently solved, as demonstrated in [17]. A similar frontier of efficiency was found by using separation rules instead of division rules in [18,19], but in this case, the frontier is from passing of communication rules of length at most two to length at most three, instead of passing from one to two. Some results about their relative environmentless counterparts were demonstrated in [20,21]. Symport/antiport rules were first introduced in tissue-like membrane systems, but later used in cell-like membrane systems, where results were surprisingly similar [22–27], giving a glance of the similarity of using both tree-like and directed graph structures.

In [28], tissue P systems with evolutional symport/antiport rules were introduced, including in communication rules the capability to evolve the objects while traveling from one region to another one. In this type of P system, two different definitions of length can be cited: On the one hand, the length of an evolutional communication rule can be defined with a single number that is related with the number of objects in the whole rule (e.g., in the rule $\left[u \, [v]_j \right]_i \rightarrow \left[v' \, [u']_j \right]_i$ is $|u| + |v| + |u'| + |v'|$); on the other hand, the length can be defined as a pair of numbers concerning the number of objects in the left-hand side and the right-hand side of the evolutional communication rules of the system (e.g., the length of the rule $\left[u \, [v]_j \right]_i \rightarrow \left[v' \, [u']_j \right]_i$ is $(|u| + |v|, |u'| + |v'|)$). In [29], several new results were presented, publishing some improvements from [28,30–32]. More precisely, in [32] an efficient family of tissue P systems with evolutional communication rules of length at most $(2, 1)$ and division rules using the environment as an active agent of the system solving the SAT problem is presented. In this work, we investigate the role of evolutional communication rules in cell-like membrane systems that use division rules as exponential workspace-generating rules, and letting the environment as a mere agent that only receives the corresponding answer of the system.

The rest of the paper is structured as follows: Section 2 is dedicated to introducing some terms used throughout the work. In the next section, cell P systems with evolutional symport/antiport rules and division rules are defined, and their recognizer versions are introduced. Sections 4 and 5 are devoted to present a solution to the problem SAT by means of a family of P systems with evolutional communication rules and division rules of length at most $(2, 1)$, and to prove the correctness of the solution. Finally, in Section 6 the results of this paper are discussed, and comparatives with other classes of P systems are given, and some open research lines are proposed, besides a description of the work in progress.

2. Preliminaries

In this section, some concepts that will be used throughout the paper will be defined. The reader can find expanded and deeper information about formal languages and membrane computing in [8,33].

An alphabet is a non-empty (finite) set, whose elements are usually called *symbols*. A *string* over Γ is a *ordered finite succession* of elements from Γ. We denote the empty string by λ.

Given two sets A and B, the *relative complement* $A \setminus B$ is defined as $A \setminus B = \{x \in A \mid x \notin B\}$. For each set A, we denote by $|A|$ the *cardinal* (number of elements) from A.

A multiset can be described explicitly as follows: $\{(a_1, \mathcal{M}(a_1)), \ldots, (a_n, \mathcal{M}(a_n))\}$, and the notation $\mathcal{M} = a_1^{\mathcal{M}(a_1)} \ldots a_n^{\mathcal{M}(a_n)}$ will be used. The cardinal of a multiset $\mathcal{M}$ over $\Gamma = \{a_1, \ldots, a_n\}$ is defined as $|\mathcal{M}| = \mathcal{M}(a_1) + \ldots + \mathcal{M}(a_n)$. We denote by $M_f(\Gamma)$ the set of all the finite multisets over Γ, and $M_f^+(\Gamma) = M_f(\Gamma) \setminus \{\emptyset\}$

Given two multisets M_1 and M_2 over Γ, the *union* of the multisets, denoted as $M_1 \cup M_2$ or $M_1 + M_2$, is the application over Γ defined as: for each $a \in \Gamma$, $(M_1 \cup M_2)(a) = M_1(a) + M_2(a)$. A multiset M_1 is included in M_2, and it is denoted by $M_1 \subseteq M_2$, if $M_1(a) \leq M_2(a)$ for each $a \in \Gamma$.

3. P Systems with Evolutional Communication and Division Rules

In this section, the framework of cell-like membrane systems where evolutional communication rules and division rules are used is introduced.

Definition 1. *A P system with evolutional symport/antiport rules and division rules of degree $q \geq 1$ is a tuple*

$$\Pi = (\Gamma, \mathcal{E}, \mu, \mathcal{M}_1, \ldots, \mathcal{M}_q, \mathcal{R}_1, \ldots, \mathcal{R}_q, i_{out})$$

where:

- Γ *is a finite (working) alphabet;*
- $\mathcal{E} \subseteq \Gamma$ *is the environment alphabet;*
- μ *is a rooted tree structure;*
- $\mathcal{M}_1, \ldots, \mathcal{M}_q$ *are the initial multisets of the membranes;*
- $\mathcal{R}_1, \ldots, \mathcal{R}_q$ *are the rules of the membranes of the system of the following form:*
  - $\left[u \left[\, \right]_j \right]_i \to \left[\left[u' \right]_j \right]_i$, *with* $0 \leq i, j \leq q, i \neq j, u \in M_f^+(\Gamma), u' \in M_f(\Gamma)$; *in the case that* $i = 0$, *then u must contain at least one object from* $\Gamma \setminus \mathcal{E}$ *(evolutional send-in symport rules);*
  - $\left[\left[u \right]_j \right]_i \to \left[u' \left[\, \right]_j \right]_i$, *with* $0 \leq i, j \leq q, i \neq j, u \in M_f^+(\Gamma), u' \in M_f(\Gamma)$ *(evolutional send-out symport rules);*
  - $\left[u \left[v \right]_j \right]_i \to \left[v' \left[u' \right]_j \right]_i$, *with* $0 \leq i, j \leq q, i \neq j, u, v \in M_f^+(\Gamma), u', v' \in M_f(\Gamma)$ *(evolutional antiport rules);*
  - $[a]_i \to [b]_i [c]_i$, *with* $1 \leq i \leq q, i \notin \{i_{out}, i_{skin}\}, a, b, c \in \Gamma$, *being* i_{skin} *the label of the skin membrane (division rules).*
- $i_{out} \in \{0, 1, \ldots, q\}$ *is the output region of the system.*

A P system with evolutional symport/antiport rules and division rules of degree $q \geq 1$ can be seen as a set of q membranes biyectively labelled by $1, \ldots, q$ organized in the rooted tree structure μ, whose root node is the *skin* membrane, and such that (a) $\mathcal{E}$ represents the set of objects that are situated in the *environment* an arbitrary number of times; (b) $\mathcal{M}_1, \ldots, \mathcal{M}_q$ represents the initial multisets situated in the q membranes of the system; (c) i_{out} is a distinguished *region* i that will encode the output of the system. The term region i will be used to denote the membrane i, in case $1 \leq i \leq q$, and to the environment in the case $i = 0$ or $i = env$. We will use env and 0 indistinguishably as the label of the environment. If $\mathcal{E} = \emptyset$, it is usually omitted from the tuple.

A *configuration* at an instant t of such a system is described by the membrane structure of the P system at the instant t, the multisets of objects from Γ in each membrane at that instant and the multiset of objects from $\Gamma \setminus \mathcal{E}$ situated in the environment. The *initial configuration* of $\Pi = (\Gamma, \mathcal{E}, \mu, \mathcal{M}_1, \ldots, \mathcal{M}_q, \mathcal{R}_1, \ldots, \mathcal{R}_q, i_{out})$ is $(\mu, \mathcal{M}_1, \ldots, \mathcal{M}_q, \emptyset)$.

A *evolutional send-in symport rule* is applicable to a configuration $\mathcal{C}_t$ at an instant t if there exists a region labelled by i that has at least a child membrane labelled by j and that it contains the multiset u of objects. The execution of such a rule $\left[u\left[\,\right]_j\right]_i \rightarrow \left[\left[u'\right]_j\right]_i$ in C_t consumes the objects of u from such a region i, and it produces the multiset of objects u' in the child membrane j in C_t.

A *evolutional send-out symport rule* is applicable to a configuration C_t at an instant t if there exists a region labelled by i that has at least a child membrane labelled by j and the child membrane contains the multiset u of objects. The execution of such a rule $\left[\left[u\right]_j\right]_i \rightarrow \left[u'\left[\,\right]_j\right]_i$ in C_t consumes the objects of u from such a region j, and it produces the multiset of objects u' in the child membrane i in C_t.

A *evolutional antiport rule* is applicable to a configuration C_t at an instant t if there exists a region labelled by i that contains a multiset of objects u and it has at least a child membrane labelled by j and the child membrane contains the multiset v of objects. The execution of such a rule $\left[u\left[v\right]_j\right]_i \rightarrow \left[v'\left[u'\right]_j\right]_i$ in C_t consumes the objects of u from such a region j and objects from v from such a region i, and it produces the multiset of objects u' in that membrane i and the multiset of objects v' in that membrane j in C_t.

A *division rule* is applicable to a configuration C_t at an instant t if there exists a region labelled by i that contains an object a. The execution of such a rule $[\,a\,]_i \rightarrow [\,b\,]_i\,[\,c\,]_i$ in C_t consumes the object a from the membrane i, the membrane is duplicated with its contents included, and objects b and c are produced one in each new membrane created in C_t.

As in tissue P systems with evolutional symport/antiport rules, two definitions of *length* (or size) can be described. On the one hand, the *length* of a evolutional communication rule $r \equiv \left[u\left[v\right]_j\right]_i \rightarrow \left[v'\left[u'\right]_j\right]_i$ can be $length(r) = (|u| + |v| + |u'| + |v'|)$; on the other hand, it can be described as a pair $length'(r) = (|u| + |v|, |u'| + |v'|)$. The first description corresponds with the sum of the cardinalities of all the multisets in the rule, while the second definition corresponds with the pair such that the first component corresponds with the sum of cardinalities of the left-hand side of the rule (LHS), and the second component corresponds with the sum of the cardinalities of the right-hand side of the rule (RHS). If r is a symport rule, then $|v| = |v'| = 0$ or $|u| = |u'| = 0$.

We say that a configuration C_t of a P system with evolutional communication rules and division rules Π produces a configuration C_{t+1} in a *transition step*, denoted by $C_t \Rightarrow_\Pi C_{t+1}$ and we say that C_{t+1} is a next configuration of C_t if we can pass from C_t to C_{t+1} applying the rules of Π according to the following principles:

- At most one rule can be applied to each object of each membrane (selected in a non-deterministic way).
- For each membrane i, either evolutional communication rules $\left[u\left[v\right]_j\right]_i \rightarrow \left[v'\left[u'\right]_j\right]_i$ or a single division rule $[\,a\,]_i \rightarrow [\,b\,]_i\,[\,c\,]_i$ can be applied. In the case of applying evolutional communication rules at configuration C_t, they would be selected in a *non-deterministic, parallel* and *maximal* way; that is, all the objects in the membrane i that can fire a rule, will fire it. In the case that a division rule $[\,a\,]_i \rightarrow [\,b\,]_i\,[\,c\,]_i$ is applied, it will be selected in a non-deterministic way. It can be seen as the division rule *blocks* the communication of the membrane with its corresponding parent membrane. All the rules are applied in an atomic way, but in order to be precise, it can be that two microsteps are executed: first, all the evolutional communication rules are executed, and then division rules are executed. It is worth taking this into account since a membrane i being divided can still communicate with its inner membranes in the same transition step.

A configuration C_t is a *halting configuration* if no rules can be applied to such configuration at an instant t.

A computation $\mathcal{C}$ of a P system Π can be described as a tuple $\mathcal{C} = (C_0, C_1, \ldots)$, where each configuration C_{t+1} can be obtained from C_t, except for the initial configuration C_0. We say that $\mathcal{C}$ is a *halting computation* of length $n+1$ if the configuration C_n is a halting configuration.

Recognizer membrane systems were introduced in [14] as a way to solve decision problems. We define here recognizer P systems with evolutional communication rules and division rules.

Definition 2. *A recognizer P system with evolutional communication rules and division rules of degree* $q \geq 1$ *is a tuple* (Π, Σ, i_{in})*, where:*

- $\Pi = (\Gamma, \mathcal{E}, \mu, \mathcal{M}_1, \ldots, \mathcal{M}_q, \mathcal{R}_1, \ldots, \mathcal{R}_q, i_{out})$ *is a P system with evolutional communication rules and division rules of degree* q*, with* $\{\texttt{yes}, \texttt{no}\} \subseteq \Gamma \setminus \mathcal{E}$ *two distinguished objects that will represent the* output *of the system, and* $\mathcal{M}_i \subseteq \Gamma \setminus \Sigma$ *for* $1 \leq i \leq q$*;*
- $\Sigma \subseteq \Gamma \setminus \mathcal{E}$ *is the input alphabet;*
- $i_{in} \in \{1, \ldots, q\}$ *is the label of the input membrane;*
- $i_{out} = 0$*;*
- *All the computations of* Π *halt.*
- *If* $\mathcal{C}$ *is a computation of* Π*, then either an object* yes *or an object* no*, but not both, are sent to the environment in the last step of the computation.*

Let $\Pi = (\Gamma, \Sigma, \mathcal{E}, \mu, \mathcal{M}_1, \ldots, \mathcal{M}_q, \mathcal{R}_1, \ldots, \mathcal{R}_q, i_{in}, i_{out})$ be a recognizer P system of this type, then for each input multiset m over Σ, we consider the system Π with input multiset m, and we denote it by $\Pi + m$. This is characterized by the fact that the multisets associated with the initial configuration of the system is: $(\mathcal{M}_1, \ldots, \mathcal{M}_{i_{in}} + m, \ldots, \mathcal{M}_q, \emptyset)$; that is, it is obtained from the initial configuration $(\mathcal{M}_1, \ldots, \mathcal{M}_{i_{in}}, \ldots, \mathcal{M}_q, \emptyset)$ of Π, by adding the multiset m to $\mathcal{M}_{i_{in}}$.

We define then a $Output(\mathcal{C})$ function whose domain is the set of computations of Π. Such a function will formalize the results or outputs of the system. If $\mathcal{C}$ is a computation of the system Π, then $Output(\mathcal{C}) = \texttt{yes}$ (respectively, $Output() = \texttt{no}$) if the object yes (resp., object no) appears in the environment associated to the *halting configuration* of $\mathcal{C}$, but does not appear in any other configuration of $\mathcal{C}$. A computation $\mathcal{C}$ will be called an *accepting computation* (respectively, a *rejecting computation*) if $Output(\mathcal{C}) = \texttt{yes}$ (resp., $Output(\mathcal{C}) = \texttt{no}$). Computations of recognizer membrane systems always halt, and will return either yes or no as a response. A recognizer membrane system with input multiset $\Pi + m$ is *confluent*, in the sense that all the computations give the same answer; that is, given an input multiset m, all the computations of a recognizer membrane system with input multiset $\Pi + m$ will be either accepting computations or rejecting computations.

The class of recognizer P systems with evolutional symport/antiport rules of length at most k (respectively, with length at most (k_1, k_2)) and division rules is denoted by $\mathcal{CDEC}((k))$ (resp., $\mathcal{CDEC}((k_1, k_2))$). If the environment does not play an active role, we say that it is a recognizer P system with symport/antiport rules of length at most k (resp., with length at most (k_1, k_2), i.e. LHS with length at most k_1 and RHS with length at most k_2) and division rules without environment, and we denote this class of systems by $\widehat{\mathcal{CDEC}}(k)$ (resp., $\widehat{\mathcal{CDEC}}(k_1, k_2)$).

In this work, a family of recognizer P systems will be used in order to solve decision problems. Let $X = (I_X, \theta_X)$ a decision problem. We say that X is solvable in polynomial time by means of a uniform family of recognizer P systems $\mathbf{\Pi} = \{\Pi(n) : n \in \mathbb{N}\}$ of the class $\mathcal{R}$, and we denote it by $X \in \mathbf{PMC}_{\mathcal{R}}$, if the following conditions hold:

(a) $\mathbf{\Pi}$ is polynomially uniform by Turing machines; that is, there exists a Turing machine that constructs $\Pi(n)$ in polynomial time.

(b) There exists a polynomial encoding (cod, s) of X such that:

  b.1 $\mathbf{\Pi}$ is polynomially bounded with respect to (X, cod, s); that is, there exists $k \in \mathbb{N}$ such that for each $u \in I_X$, each computation of $\Pi(s(u)) + cod(u)$ runs, at most, for $|u|^k$ computation steps.

  b.2 $\mathbf{\Pi}$ is sound and complete with respect to (X, cod, s).

4. Methods

Let φ a propositional formula in CNF with n variables and p clauses. Let $s(\varphi) = \langle n, p\rangle$ and $cod(\varphi)$ contains $x_{i,j,0}$ if the literal x_i is in the clause C_j, $\overline{x}_{i,j,0}$ if the literal $\neg x_i$ is in the clause C_j and $x^*_{i,j,0}$ if the variable x_i does not appear in the clause C_j. Let the function $\perp(i,j) = i \perp j = i + jn$. Then, the system $\Pi(\langle n,p\rangle)$ will be the responsible of solving φ. Let

$$\Pi(\langle n,p\rangle) = (\Gamma, \Sigma, \mathcal{E}, \mu, \mathcal{M}_1, \dots, \mathcal{M}_{np+5}, \mathcal{R}_1, \dots, \mathcal{R}_{np+5}, i_{in}, i_{out})$$

a recognizer membrane system from $\mathcal{CDEC}((2,1))$, where:

1. The working alphabet $\Gamma = \Sigma \cup \{\mathtt{yes}, \mathtt{no}, y_1, y_2, n_1, n_2, \alpha', \#\} \cup$ $\{a_{i,j}, T_{i,j}, F_{i,j}, x_{i,j}, \overline{x}_{i,j}, x^*_{i,j} \mid 1 \le i \le n, 1 \le j \le p\} \cup$ $\{x_{i,j,k}, \overline{x}_{i,j,k}, x^*_{i,j,k} \mid 1 \le i \le n, 1 \le j \le p, 1 \le k \le np + \lfloor np/2 \rfloor + 1\} \cup$ $\{c_j \mid 1 \le j \le p\} \cup \{\alpha_j \mid 1 \le j \le p+1\} \cup \{\gamma_k \mid 0 \le k \le np+2\} \cup$ $\{\delta_k \mid 0 \le k \le np+3\} \cup \{\delta'_k \mid 0 \le k \le np+1\}$.
2. The input alphabet $\Sigma = \{x_{i,j,0}, \overline{x}_{i,j,0}, x^*_{i,j,0} \mid 1 \le i \le n, 1 \le j \le p\}$.
3. Environment alphabet $\mathcal{E} = \emptyset$.
4. $\mu = [[\ ]_1 [\ ]_2 \cdots [\ ]_{np} [\ ]_{np+2} [\ ]_{np+3} [\ ]_{np+4} [\ ]_{np+5}]_{np+1}$
5. $\mathcal{M}_k = \emptyset, 1 \le k \le np+1$,
   $\mathcal{M}_{np+2} = \{a_{i,j} \mid 1 \le i \le n, 1 \le j \le p\} \cup \{\alpha_j \mid 1 \le j \le p+1\}$,
   $\mathcal{M}_{np+3} = \{\delta'_0\}$, $\mathcal{M}_{np+4} = \{\delta_0\}$, $\mathcal{M}_{np+5} = \{\gamma_0\}$.
6. The set of rules $\mathcal{R} = \mathcal{R}_1 \cup \dots \mathcal{R}_{np+5}$ contains the following rules:

   6.1 Rules for generating all the necessary γ_{np+2} to simulate the environment.

   $$[\gamma_k]_{np+5} \to [\gamma_{k+1}]_{np+5} [\gamma_{k+1}]_{np+5} \quad \text{for} \quad 0 \le k \le np+1$$
   $$\big[[\gamma_{np+2}]_{np+5} \big]_{np+1} \to \big[\gamma_{np+2} [\ ]_{np+5} \big]_{np+1}$$

   6.2 Rules to generate p copies of the 2^n possible truth assignments. For that, 2^{np} "partial" truth assignments will be generated.

   $$\left.\begin{array}{l} [a_{i,j}]_{np+2} \to [T_{i,j}]_{np+2} [F_{i,j}]_{np+2} \\ \big[[T_{i,j} F_{i,j'}]_{np+2} \big]_{np+1} \to \big[\# [\ ]_{np+2} \big]_{np+1} \end{array}\right\} \text{for} \quad \begin{array}{l} 1 \le i \le n \\ 1 \le j, j', \le p \end{array}$$

   6.3 Rules to generate 2^{np} copies of $cod(\varphi)$.

   $$\left.\begin{array}{l} \big[x_{i,j,0} [\ ]_{i \perp j} \big]_{np+1} \to \big[[x_{i,j,1}]_{i \perp j} \big]_{np+1} \\ \big[\overline{x}_{i,j,0} [\ ]_{i \perp j} \big]_{np+1} \to \big[[\overline{x}_{i,j,1}]_{i \perp j} \big]_{np+1} \\ \big[x^*_{i,j,0} [\ ]_{i \perp j} \big]_{np+1} \to \big[[x^*_{i,j,1}]_{i \perp j} \big]_{np+1} \end{array}\right\} \text{for} \quad 1 \le i \le n, 1 \le j \le p$$

   $$\left.\begin{array}{l} \big[x_{i,j,k} \big]_{i \perp j} \to [x_{i,j,k+1}]_{i \perp j} [x_{i,j,k+1}]_{i \perp j} \\ \big[\overline{x}_{i,j,k} \big]_{i \perp j} \to [\overline{x}_{i,j,k+1}]_{i \perp j} [\overline{x}_{i,j,k+1}]_{i \perp j} \\ \big[x^*_{i,j,k} \big]_{i \perp j} \to [x^*_{i,j,k+1}]_{i \perp j} [x^*_{i,j,k+1}]_{i \perp j} \end{array}\right\} \text{for} \quad \begin{array}{l} 1 \le i \le n \\ 1 \le j \le p \\ 1 \le k \le np + \lfloor np/2 \rfloor \end{array}$$

   $$\left.\begin{array}{l} \big[[x_{i,j,np+\lfloor np/2 \rfloor+1}]_{i \perp j} \big]_{np+1} \to \big[x_{i,j} [\ ]_{i \perp j} \big]_{np+1} \\ \big[[\overline{x}_{i,j,np+\lfloor np/2 \rfloor+1}]_{i \perp j} \big]_{np+1} \to \big[\overline{x}_{i,j} [\ ]_{i \perp j} \big]_{np+1} \\ \big[[x^*_{i,j,np+\lfloor np/2 \rfloor+1}]_{i \perp j} \big]_{np+1} \to \big[x^*_{i,j} [\ ]_{i \perp j} \big]_{np+1} \end{array}\right\} \text{for} \quad 1 \le i \le n, 1 \le j \le p$$

   $$\big[\gamma_{np+2} [\delta'_0]_{np+3} \big]_{np+1} \to \big[[\delta'_1]_{np+3} \big]_{np+1}$$
   $$[\delta'_k]_{np+3} \to [\delta'_{k+1}]_{np+3} [\delta'_{k+1}]_{np+3} \quad \text{for } 1 \le k \le np$$
   $$\big[[\delta'_{np+1}]_{np+3} \big]_{np+1} \to \big[\delta'_{np+1} [\ ]_{np+3} \big]_{np+1}$$

   6.4 Rules to check which clauses are satisfied.

$$\left.\begin{array}{l}\left[x_{i,j}\,[\,T_{i,j}\,]_{np+2}\right]_{np+1} \rightarrow \left[\,[\,c_j\,]_{np+2}\right]_{np+1}\\ \left[\overline{x}_{i,j}\,[\,T_{i,j}\,]_{np+2}\right]_{np+1} \rightarrow \left[\,[\,\#\,]_{np+2}\right]_{np+1}\\ \left[x^*_{i,j}\,[\,T_{i,j}\,]_{np+2}\right]_{np+1} \rightarrow \left[\,[\,\#\,]_{np+2}\right]_{np+1}\\ \left[x_{i,j}\,[\,F_{i,j}\,]_{np+2}\right]_{np+1} \rightarrow \left[\,[\,\#\,]_{np+2}\right]_{np+1}\\ \left[\overline{x}_{i,j}\,[\,F_{i,j}\,]_{np+2}\right]_{np+1} \rightarrow \left[\,[\,c_j\,]_{np+2}\right]_{np+1}\\ \left[x^*_{i,j}\,[\,F_{i,j}\,]_{np+2}\right]_{np+1} \rightarrow \left[\,[\,\#\,]_{np+2}\right]_{np+1}\end{array}\right\}\text{ for } 1 \le i \le n, 1 \le j \le p$$

6.5 Rules to check if all the clauses are satisfied by a truth assignment.

$$\left[\delta'_{np+1}\,[\,\alpha_{p+1}\,]_{np+2}\right]_{np+1} \rightarrow \left[\,[\,\alpha'\,]_{np+2}\right]_{np+1}$$
$$\left[\,[\,c_j\alpha_j\,]_{np+2}\right]_{np+1} \rightarrow \left[\#\,[\,]_{np+2}\right]_{np+1} \quad \text{for } 1 \le j \le p$$
$$\left[\,[\,\alpha_j\alpha'\,]_{np+2}\right]_{np+1} \rightarrow \left[n_1\,[\,]_{np+2}\right]_{np+1} \quad \text{for } 1 \le j \le p$$

6.6 General counter.

$$\left[\gamma_{np+2}\,[\,\delta_k\,]_{np+5}\right]_{np+1} \rightarrow \left[\,[\,\delta_{k+1}\,]_{np+5}\right]_{np+1} \quad \text{for } 0 \le k \le np+2$$
$$\left[\,[\,\delta_{np+3}\,]_{np+5}\right]_{np+1} \rightarrow \left[\delta_{np+3}\,[\,]_{np+5}\right]_{np+1}$$

6.7 Rules to return a negative answer.

$$\left[n_1\,[\,]_{np+2}\right]_{np+1} \rightarrow \left[\,[\,n_1\,]_{np+2}\right]_{np+1}$$
$$\left[\delta_{np+3}\,[\,n_1\,]_{np+2}\right]_{np+1} \rightarrow \left[n_2\,[\,]_{np+2}\right]_{np+1}$$
$$\left[\,[\,n_2\,]_{np+1}\right]_0 \rightarrow \left[\texttt{no}\,[\,]_{np+1}\right]_0$$

6.8 Rules to return an affirmative answer.

$$\left[\delta_{np+3}\,[\,\alpha'\,]_{np+2}\right]_{np+1} \rightarrow \left[\,[\,y_1\,]_{np+2}\right]_{np+1}$$
$$\left[\,[\,y_1\,]_{np+2}\right]_{np+1} \rightarrow \left[y_2\,[\,]_{np+2}\right]_{np+1}$$
$$\left[\,[\,y_2\,]_{np+1}\right]_0 \rightarrow \left[\texttt{yes}\,[\,]_{np+1}\right]_0$$

6.9 Input membrane $i_{in} = np+1$ and output membrane $i_{out} = 0$.

In this section, the behaviour of a recognizer P system from $\mathcal{CDEC}((2,1))$ solving an instance φ from SAT is described. Let $\varphi = C_1 \wedge \ldots \wedge C_p$ be a propositional logic formula in CNF, where $C_j = l_1 \vee \ldots \vee l_{r_j}, l_k \in \{x_i, \neg x_i \mid 1 \le i \le n\}$. Then φ will be processed by the system $\Pi(s(\varphi)) + cod(\varphi)$, where $s(\varphi) = \langle n, p \rangle$ and $cod(\varphi) = \{x_{i,j,0} \mid x_i \in C_j\} \cup \{\overline{x}_{i,j,0} \mid \neg x_i \in C_j\} \cup \{x^*_{i,j,0} \mid x_i \notin C_j \wedge \neg x_i \notin C_j\}$.

Let us note $cod_k(\varphi)$ the set of all the elements from $cod(\varphi)$ with the third subscript equal to k. Let us note $cod^*(\varphi)$ the set of all the elements from $cod(\varphi)$ without the third subscript.

The solution follows a brute force algorithm protocol in the framework of recognizer P systems with evolutional symport/antiport rules and division rules, and consists of the following stages:

4.1. Generation Stage

In the generation stage, different elements necessary for the rest of the computation are generated at the same time. First, in order to avoid the use of the environment to obtain γ objects, it is necessary to use some of the synchronization protocols in [32], 2^{np+2} copies of the object γ_{np+2} will be generated in membranes $np+5$ through the rules from 6.1. These objects will be present in the membrane $np+1$ at configuration $\mathcal{C}_{np+3}$. For generating the different truth assignments, objects $T_{i,j}$ and $F_{i,j}$ will represent a partial truth assignment of the variable x_i in the following sense: Objects $T_{i,j}$ and $F_{i,j'}$, with $i \neq j$, would be incompatible, since two different values would be assigned to the same variable. Therefore, in order to remove this possibility, these two objects will be removed from the system by means of the rule $\left[\,[\,T_{i,j}F_{i,j'}\,]_{np+2}\,\right]_{np+1} \rightarrow \left[\#\,[\,]_{np+2}\,\right]_{np+1}$. After applying these rules, only "real" truth assignments would be present in the system. In fact, there can be assignments where a variables has not been assigned any value. In this case, it will not return a false positive. For generating the different np partial truth assignments, np computational steps should be applied, but since rules are fired in a non-deterministic way, and the removal of these incompatible variables could be applied in between the process of generation, at most $\lfloor np/2 \rfloor$ extra steps should be taken into account. Therefore, in configuration $\mathcal{C}_{np+\lfloor np/2 \rfloor}$, membranes labelled by $np+2$ will contain all the possible truth assignments.

Besides, objects from $cod(\varphi)$ will be sent into their corresponding membrane. From the second computational step, membranes $1,\ldots,np$ will be duplicated in each step until having $2^{np+\lfloor np/2 \rfloor}$ copies of the corresponding object from $cod_{np+\lfloor np/2 \rfloor}(\varphi)$. These objects will be sent out to the membrane $np+1$, and therefore in the configuration $\mathcal{C}_{np+\lfloor np/2 \rfloor+1}$, $2^{np+\lfloor np/2 \rfloor}$ copies of $cod^*(\varphi)$ will be present in the membrane $np+1$. In this configuration, the next stage begins.

4.2. First Checking Stage

In this stage, objects from $cod^*(\varphi)$ will react with objects $T_{i,j}$ and $F_{i,j}$ through rules from 6.4. In this stage, an object c_j will be generated in a membrane $np+2$ if and only if the truth assignment represented in that membrane makes true the corresponding literal. This stage takes one computational step. At the same time, counters δ_k and δ'_k are being generated. δ_k will evolve by using objects γ_{np+2} as "catalysts", and np copies of the object δ'_{np+1} will be present at the membrane $np+1$ at configuration $\mathcal{C}_{2np+5}$. When this happens, the second checking stage starts.

4.3. Second Checking Stage

The $(2np+6)$-th step will consist of the application of the rule $\left[\,\delta'_{np+1}\,[\,\alpha_{p+1}\,]_{np+2}\,\right]_{np+1} \rightarrow \left[\,[\,\alpha'\,]_{np+2}\,\right]_{np+1}$, creating an object α' in each membrane labelled by $np+2$. At the same time, objects c_j present in a membrane will remove the corresponding object a_j; that is, the absence of an object a_j in a membrane $np+2$ represents that clause C_j is satisfied by the corresponding truth assignment. The object α' will fire a rule $\left[\,[\,\alpha_j\alpha'\,]_{np+2}\,\right]_{np+1} \rightarrow \left[\,n_1\,[\,]_{np+2}\,\right]_{np+1}$ if there exists an object a_j in such a membrane. This stage takes exactly two computational steps.

4.4. Output Stage

In configuration $\mathcal{C}_{2np+7}$, the existence of an object α' in a membrane labelled by $np+2$ implies that no objects α_j remained in such a membrane; that is, that all clauses are satisfied by the corresponding truth assignment. Therefore, if there exists an object α' in such a membrane, it implies that the formula φ is satisfiable. Thus, two different scenarios can be observed. On the one hand, if the formula is satisfiable, there exists at least one membrane labelled by $np+2$ at configuration $\mathcal{C}_{2np+7}$ such that it contains an object α'. In the next step, an object y_1 will be generated in such a membrane, that will be sent out, first to membrane $np+1$ as an object y_2, and finally to the environment as an object `yes`. On the other hand, if the formula is unsatisfiable, no objects α' remain in any of the membranes labelled by $np+2$ at configuration $\mathcal{C}_{2np+7}$. Objects n_1 will be sent to the membrane $np+1$ and, in the next step, they will be sent back to any membrane $np+2$, taking into account that the target membranes are selected in a non-deterministic way. In the next step, as the object δ_{np+3} still exists in membrane $np+1$, it reacts with an object n_1, transforming it into an object n_2 at the skin membrane, and in the last step of the configuration, it will be sent out to the environment as an object `no`. It is important to take into account that, in the affirmative case, object δ_{np+3} is consumed, and since only one object of this kind exists, objects n_1 will not have any objects to react with. This stage takes exactly three computational steps, both in the affirmative case and in the negative case. In Figure 1, a graphical description of this process is provided to clarify how this stage works.

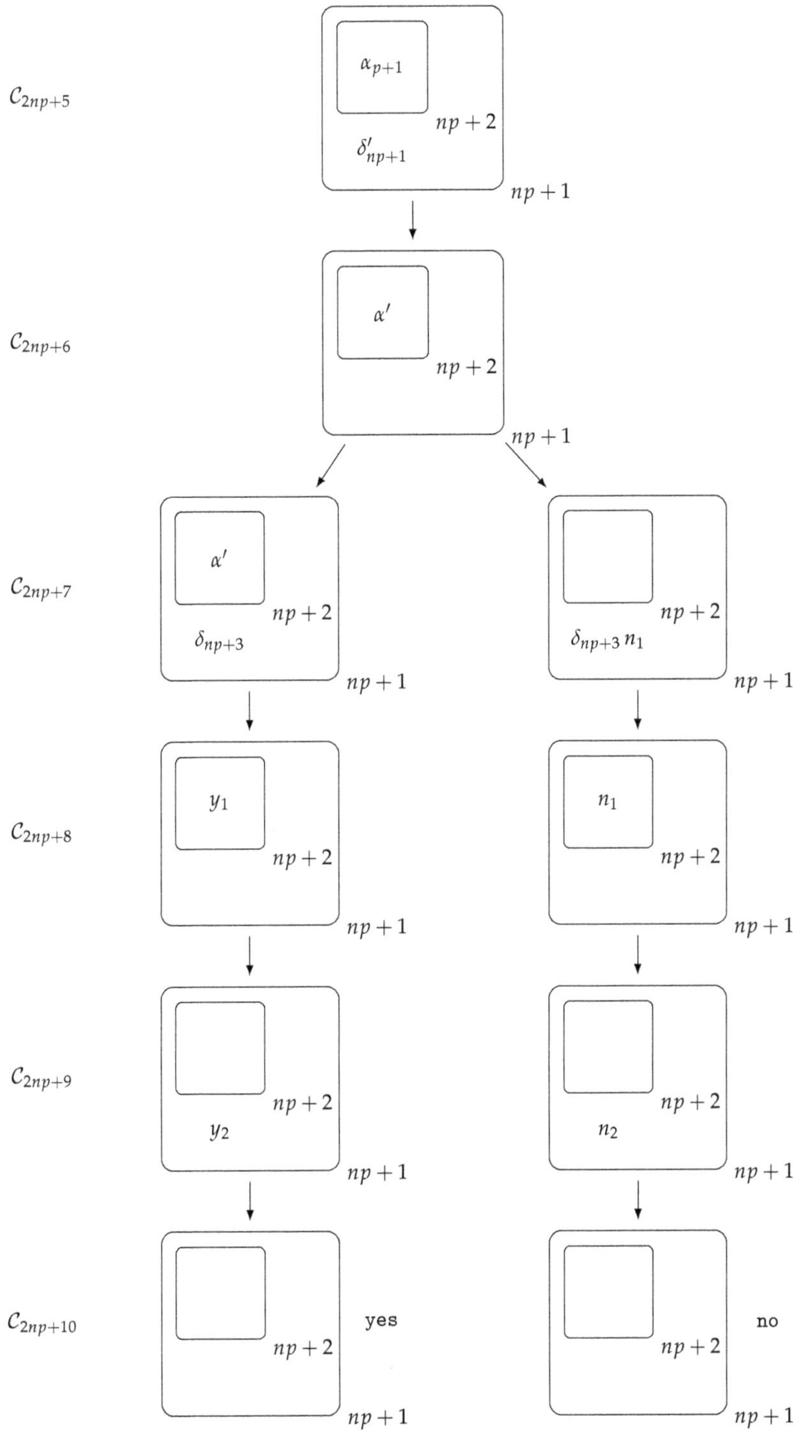

Figure 1. Evolution of the final stage in the affirmative case (**left**) and in the negative case (**right**).

5. Results

The results of the paper will be discussed in this section.

Theorem 1. $\mathtt{SAT} \in \mathbf{PMC}_{\mathcal{CDEC}((2,1))}$

Proof. The family of P systems constructed previously verifies the following:

- All the systems from $\mathbf{\Pi} = \{\Pi(n) \mid n \in \mathbb{N}\}$ are recognizer P systems from $\mathcal{CDEC}((2,1))$.
- The family $\mathbf{\Pi}$ is polynomially uniform by Turing machines given that, for each $n, p \in \mathbb{N}$, rules of $\Pi(\langle n, p\rangle)$ from $\mathbf{\Pi}$ are defined recursively by $n, p \in \mathbb{N}$, and the quantity of resources needed for constructing an element of the family is of polynomial order with respect to n and p, as it is shown below:
  - Alphabet size: $3n^2p^2 + (11 + \lfloor \frac{np}{2} \rfloor)np + 2p + 16 \in \Theta(n^2p^2)$
  - Initial number of membranes: $np + 5 \in \Theta(np)$
  - Initial number of objects in cells: $np + p + 4 \in \Theta(np)$
  - Number of rules: $3n^2p^2\lfloor \frac{np}{2} \rfloor + 2np^2 + 15np + 2p + 16 \in \Theta(n^3p^3)$
  - Maximum number of objects involved in a rule: $3 \in \Theta(1)$.
- The pair of polynomial-time computable functions (cod, s) defined complies the following: for each formula φ from $\mathtt{SAT}$, $s(\varphi)$ is a natural number, $cod(\varphi)$ is the input multiset of the system $\Pi(s(\varphi))$ and for each $t \in \mathbb{N}$, $s^{-1}(t)$ is a finite set.
- The family $\mathbf{\Pi}$ is polynomially bounded in time: in fact, for each formula φ from $\mathtt{SAT}$, the recognizer P system $\Pi(s(\varphi)) + cod(\varphi)$ takes exactly $2np + 10$ computational steps in return an answer, either positive or negative, being n the number of variables and p the number of clauses of φ.
- The family $\mathbf{\Pi}$ is sound with respect to (X, cod, s): in fact, for each formula φ, if the computations of $\Pi(s(\varphi)) + cod(\varphi)$ are *accepting computations*, then φ is satisfiable.
- The family $\mathbf{\Pi}$ is complete with respect to (X, cod, s): in fact, for each formula φ that is satisfiable, all the computations of $\Pi(s(\varphi)) + cod(\varphi)$ are *accepting computations*.

□

Corollary 1. $\mathbf{NP} \cup \mathbf{co-NP} \subseteq \mathbf{PMC}_{\mathcal{CDEC}((2,1))}$

Proof. It is enough to observe that $\mathtt{SAT}$ is an **NP**-complete problem, $\mathtt{SAT} \in \mathbf{PMC}_{\mathcal{CDEC}((2,1))}$ and the class $\mathbf{PMC}_{\mathcal{CDEC}((2,1))}$ is closed under polynomial-time reduction and under complementary. □

In fact, this family does not use the environment with an active role, therefore:

Corollary 2. $\mathbf{NP} \cup \mathbf{co-NP} \subseteq \mathbf{PMC}_{\widehat{\mathcal{CDEC}}(2,1)}$

Proof. It is enough to observe that $\mathbf{NP} \cup \mathbf{co-NP} \subseteq \mathbf{PMC}_{\mathcal{CDEC}((2,1))}$ from the fact that a uniform family of recognizer membrane systems from $\mathcal{CDEC}((2,1))$ solving the problem $\mathtt{SAT}$ in polynomial time has been constructed, and this solution does use the environment only as the output of the system, $\mathcal{E} = \emptyset$, and does not take any object from it. □

Corollary 3. $\mathbf{NP} \cup \mathbf{co-NP} \subseteq \mathbf{PMC}_{\widehat{\mathcal{CDEC}}(3)}$

6. Discussion

The idea of this solution is to use the power of division rules to generate all the possible truth assignments and objects first, and later on to use the parallel communication between membranes to transport all the needed objects. It is important to take into account that this is a great difference with P systems with active membranes, since in these systems, communication rules between membranes are limited to one object per membrane and time step. While the first stage takes the majority of the time, checking of the clauses and output of the system are executed in five time steps, using the created workspace in the generation stage. This implies that a great optimization (for instance, in the form of a parallel implementation) would be needed in order to generate the exponential number of membranes in polynomial (in this case, linear) time. In a practical way, this could lead to an interesting competitor with respect to the state-of-art SAT solvers, that are necessary to solve industrial propositional logic formulae used to improve some engineering processes.

7. Contributions

In this work, a solution to SAT by means of a family of recognizer membrane systems from $\widehat{\mathcal{CDEC}}(2,1)$ is given. In previous works, a similar result in the tissue-like counterpart was given, but using the environment as an active element. An interesting work would be to prove that the role of the environment is also irrelevant in tissue P systems with evolutional symport/antiport rules and division rules. Besides, similar results using separation rules instead of division rules were provided in the same work. Taking into account the differences between division rules and separation rules and between tissue-like and cell-like, it would be interesting to see if this result can also be translated to the cell-like framework. In this sense, a complete study of the role of the environment while using separation rules, both in the tissue-like and in the cell-like frameworks will be studied.

Author Contributions: Conceptualization, D.O.-M. and M.J.P.-J.; methodology, D.O.-M. and L.V.-C.; validation, D.O.-M., L.V.-C. and M.J.P.-J.; formal analysis, M.J.P.-J.; investigation, D.O.-M.; writing—original draft preparation, D.O.-M.; writing—review and editing, L.V.-C.; supervision, L.V.-C. and M.J.P.-J. All authors have read and agreed to the published version of the manuscript.

Funding: This research was funded by "FEDER/Ministerio de Ciencia e Innovación—Agencia Estatal de Investigación/_Proyecto (TIN2017-89842-P)"—MABICAP. D.O.-M. also acknowledges Contratación de Personal Investigador Doctor. (Convocatoria 2019) 43 Contratos Capital Humano Línea 2. Paidi 2020, supported by the European Social Fund and Junta de Andalucía.

Institutional Review Board Statement: Not applicable.

Informed Consent Statement: Not applicable.

Data Availability Statement: This work is self-contained, and all the data can be found within the own paper.

Conflicts of Interest: The authors declare no conflict of interest.

References

1. Păun, G. Computing with membranes. *Turku Cent. Comput.-Sci.-Tucs Rep.* **1998**, *208*, 108–143.
2. Martín-Vide, C.; Păun, G.; Pazos, J.; Rodríguez-Patón, A. Tissue P systems. *Theor. Comput. Sci.* **2003**, *296*, 295–326. [CrossRef]
3. Ionescu, M.; Păun, G.; Yokomori, T. Spiking neural P systems. *Fundam. Inform.* **2006**, *71*, 279–308.
4. Bao, T.; Yang, Q.; Peng, H.; Luo, X.; Wang, J. Computational power of sequential dendrite P systems. *Theor. Comput. Sci.* **2021**, *893*, 133–145. [CrossRef]
5. Wang, L.; Liu, X.; Zhao, Y. Universal Nonlinear Spiking Neural P Systems with Delays and Weights on Synapses. *Comput. Intell. Neurosci.* **2021**, *2021*, 3285719. [CrossRef]
6. Wu, T.; Pan, L.; Yu, Q.; Tan, K.C. Numerical Spiking Neural P Systems. *IEEE Trans. Neural Netw. Learn. Syst.* **2021**, *32*, 2443–2457. [CrossRef] [PubMed]
7. Păun, G. *Membrane Computing: An Introduction*, 1st ed.; Springer: Berlin/Heidelberg, Germany, 2002.
8. Păun, G.; Rozenberg, G.; Salomaa, A. *The Oxford Handbook of Membrane Computing*, 1st ed.; Oxford University Press, Inc.: New York, NY, USA, 2010.
9. Ciobanu, G.; Pérez-Jiménez, M.J.; Păun, G. *Applications of Membrane Computing*, 1st ed.; Springer: Berlin/Heidelberg, Germany, 2006.
10. Frisco, P.; Gheorghe, M.; Pérez-Jiménez, M.J. *Applications of Membrane Computing in Systems and Synthetic Biology*, 1st ed.; Springer International Publishing: Basel, Switzerland, 2014.
11. Zhang, G.; Pérez-Jiménez, M.J.; Gheorghe, M. *Real-Life Applications with Membrane Computing*, 1st ed.; Springer: Cham, Switzerland, 2017.
12. Zhang, G.; Pérez-Jiménez, M.J.; Riscos-Núñez, A.; Verlan, S.; Konur, S.; Hinze, T.; Gheorghe, M. *Membrane Computing Models: Implementations*, 1st ed.; Springer: Singapore, 2021.
13. Păun, G. Computing with membranes: Attacking NP-Complete Problems. In Proceedings of the Second International Conference on Unconventional Models of Computation, Brussels, Belgium, 13–16 December 2000; pp. 94–115.

14. Pérez-Jiménez, M.J.; Romero-Jiménez, Á.; Sancho-Caparrini, F. Complexity classes in models of cellular computing with membranes. *Nat. Comput.* **2003**, *2*, 265–285. [CrossRef]
15. Păun, G.; Pérez-Jiménez, M.J.; Riscos-Núñez, A. Tissue P Systems with Cell Division. In Proceedings of the Second Brainstorming Week on Membrane Computing, Sevilla, Spain, 2–7 February 2004; pp. 380–386.
16. Porreca, A.E.; Murphy, N.; Pérez-Jiménez, M.J. An optimal frontier of the efficiency of tissue P systems with cell division. In Proceedings of the Tenth Brainstorming Week on Membrane Computing, Sevilla, Spain, 30 January–3 February 2012; pp. 141–166.
17. Díaz-Pernil, D.; Pérez-Jiménez, M.J.; Romero-Jiménez, Á. Efficient simulation of tissue-like P systems by transition cell-like P systems. *Nat. Comput.* **2009**, *8*, 797–806. [CrossRef]
18. Pan, L.; Pérez-Jiménez, M.J.; Riscos-Núñez, A.; Rius, M. New frontiers of the efficiency in tissue P systems. In Proceedings of the Asian Conference on Membrane Computing (ACMC 2012), Wuhan, China, 15–18 October 2012; pp. 61–73.
19. Pérez-Jiménez, M.J.; Sosík, P. An Optimal Frontier of the Efficiency of Tissue P Systems with Cell Separation. *Fundam. Inform.* **2015**, *138*, 45–60. [CrossRef]
20. Macías-Ramos, L.F.; Pérez-Jiménez, M.J.; Riscos-Núñez, A.; Rius, M.; Valencia-Cabrera, L. The efficiency of tissue P systems with cell separation relies on the environment. In Proceedings of the Thirteenth International Conference on Membrane Computing (CMC 2012), Budapest, Hungary, 28–31 August 2013; pp. 277–290.
21. Pérez-Jiménez, M.J.; Riscos-Núñez, A.; Rius, M.; Romero-Campero, F.J. A polynomial alternative to unbounded environment for tissue P systems with cell division. *Int. J. Comput. Math.* **2013**, *90*, 760–775. [CrossRef]
22. Macías-Ramos, L.F.; Pérez-Jiménez, M.J.; Riscos-Núñez, A.; Valencia-Cabrera, L. Membrane fission versus cell division: When membrane proliferation is not enough. *Theor. Comput. Sci.* **2015**, *608*, 57–65. [CrossRef]
23. Macías-Ramos, L.F.; Song, B.; Song, T.; Pan, L.; Pérez-Jiménez, M.J. Limits on efficient computation in P systems with symport/antiport. In Proceedings of the Fifteenth Brainstorming Week on Membrane Computing, Sevilla, Spain, 31 January–3 February 2017; pp. 147–160.
24. Macías-Ramos, L.F.; Song, B.; Valencia-Cabrera, L.; Pan, L.; Pérez-Jiménez, M.J. Membrane fission: A computational complexity perspective. *Complexity* **2016**, *21*, 321–334. [CrossRef]
25. Orellana-Martín, D.; Valencia-Cabrera, L.; Riscos-Núñez, A.; Pérez-Jiménez, M.J. A path to computational efficiency through membrane computing. *Theor. Comput. Sci.* **2019**, *777*, 443–453. [CrossRef]
26. Orellana-Martín, D.; Valencia-Cabrera, L.; Song, B.; Pan, L.; Pérez-Jiménez, M.J. P systems with symport/antiport rules: When do the surroundings matter? *Theor. Comput. Sci.* **2020**, *805*, 206–217. [CrossRef]
27. Valencia-Cabrera, L.; Song, B.; Macías-Ramos, L.F.; Pan, L.; Riscos-Núñez, A.; Pérez-Jiménez, M.J. Minimal cooperation in P systems with symport/antiport: A complexity approach. In Proceedings of the Thirteenth Brainstorming Week on Membrane Computing, Sevilla, Spain, 2–6 February 2015; pp. 301–323.
28. Song, B.; Zhang, C.; Pan, L. Tissue-like P systems with evolutional symport/antiport rules. *Inf. Sci.* **2017**, *378*, 177–193. [CrossRef]
29. Orellana-Martín, D. The P vs NP Problem: Development of New Techniques through Bio-Inspired Models of Computation. Ph.D. Thesis, University of Sevilla, Sevilla, Spain, 2019.
30. Pan, L.; Song, B.; Valencia-Cabrera, L.; Pérez-Jiménez, M.J. The computational complexity of tissue P systems with evolutional symport/antiport rules. *Complexity* **2018**, *3745210*, 21. [CrossRef]
31. Orellana-Martín, D.; Valencia-Cabrera, L.; Song, B.; Pan, L.; Pérez-Jiménez, M.J. Narrowing frontiers with evolutional communication rules and cell separation. In Proceedings of the Sixteenth Brainstorming Week on Membrane Computing, Sevilla, Spain, 30 January–2 February 2018; pp. 123–162.
32. Orellana-Martín, D.; Valencia-Cabrera, L.; Song, B.; Pan, L.; Pérez-Jiménez, M.J. Tuning Frontiers of Efficiency in Tissue P systems with Evolutional Communication Rules. *Complexity* **2021**, *2021*, 7120840. [CrossRef]
33. Rozenberg, G.; Salomaa, A. *Handbook of Formal Languages*, 1st ed.; Springer: Berlin/Heidelberg, Germany, 1997.

axioms

MDPI

Article

List Approximation for Increasing Kolmogorov Complexity

Marius Zimand

Department of Computer and Information Sciences, Towson University, Baltimore, MD 21252, USA;
mzimand@towson.edu

Abstract: It is impossible to effectively modify a string in order to increase its Kolmogorov complexity. However, is it possible to construct a few strings, no longer than the input string, so that most of them have larger complexity? We show that the answer is yes. We present an algorithm that takes as input a string x of length n and returns a list with $O(n^2)$ strings, all of length n, such that 99% of them are more complex than x, provided the complexity of x is less than $n - \log\log n - O(1)$. We also present an algorithm that obtains a list of quasi-polynomial size in which each element can be produced in polynomial time.

Keywords: Kolmogorov complexity; random strings; extractors

MSC: 68Q30

Citation: Zimand, M. List Approximation for Increasing Kolmogorov Complexity. *Axioms* **2021**, *10*, 334. https://doi.org/10.3390/axioms10040334

Academic Editor: Victor Mitrana

Received: 27 September 2021
Accepted: 30 November 2021
Published: 7 December 2021

1. Introduction

The Kolmogorov complexity of a binary string x, denoted $C(x)$, is the minimal description length of x, i.e., it is the length of the shortest program (in a fixed universal programming system) that prints x. We analyze the possibility of modifying a string in an effective way in order to obtain a string with higher complexity, without increasing its length. Strings with high complexity exhibit good randomness properties and are potentially useful, because they can be employed in lieu of random bits in probabilistic algorithms. It is common to define the randomness deficiency of x as the difference $|x| - C(x)$ (where $|x|$ is the length of x) and to say that the smaller the randomness deficiency is, the more random the string is. In this sense, we want to modify a string so that it becomes "more" random. As stated, the above task is impossible, because, clearly, any effective modification cannot increase the Kolmogorov complexity (at least not by more than a constant). If f is a computable function, $C(f(x)) \leq C(x) + O(1)$, for every x. Consequently, we have to settle for a weaker solution and the one we consider is that of list approximation. List approximation consists in the construction of a list of objects guaranteed to contain at least one element having the desired property. Here, we try to obtain a stronger type of list approximation, in which, not just one, but *most* of the elements in the list have the desired property. More precisely, we study the following question.

Question. Is there a computable function which takes as input a string x and outputs a short list of strings, which are not longer than x, such that most of the elements in the list have complexity greater than $C(x)$?

The formulation of the question rules out some trivial and non-interesting answers. First, the requirement that the list is "short" is necessary, because, otherwise, we can ignore the input x and simply take all strings of length n and most of them have complexity at least $n-2$, which is within $O(1)$ of the largest complexity of strings of length n. Secondly, the restriction that the length is not increased is also necessary, because, otherwise, we can append to the input x a random string and obtain, with high probability, a more complex string (see the discussion in Section 2). These restrictions not only make the problem interesting, but also amenable to applications in which the input string and the modified strings need to be in a given finite set. The solution that we give can be readily adjusted to handle such applications.

There are several parameters to consider. The first one is the size of the list. The shorter the list is, the better the approximation is. Next, the increasing-complexity procedure that we seek does not work for all strings x. Let us recall that $C(x) \leq |x| + O(1)$ and, if x is a string of maximal complexity at its length, then there simply is no string of larger complexity at its length. In general, for strings x that have complexity close to $|x|$, it is difficult to increase their complexity. Thus, a second parameter is the bound on the complexity of x for which the increasing-complexity procedure succeeds. The closer this bound is to $|x|$, the better the procedure is. The third parameter is the complexity of the procedure. The procedure is required to be computable, but it is preferable if it is computable in polynomial time.

We show the following two results. The first one exhibits a computable list approximation for increasing the Kolmogorov complexity that works for any x with complexity $C(x) < |x| - \log\log|x| - O(1)$.

Theorem 1 (Computable list of quadratic size for increasing the Kolmogorov complexity). *There exists a computable function f that takes as input $x \in \{0,1\}^*$ and a rational number $\delta > 0$ and returns a list of strings of length at most $|x|$ with the following properties:*

1. *The size of the list is $O(|x|^2)\mathrm{poly}(1/\delta)$;*
2. *If $C(x) < |x| - \log\log|x| - O(1)$, then the $(1-\delta)$ fraction of the elements in the list $f(x)$ have a Kolmogorov complexity larger than $C(x)$ (where the constant hidden in $O(1)$ depends on δ).*

Whether the bound $C(x) < |x| - \log\log|x| - O(1)$ can be improved remains open. Further reducing the list size is also an interesting open question. We could not establish a lower bound and, as far as we currently know, it is possible that even a constant list size may be achievable.

In the next result, the complexity-increasing procedure runs in polynomial time in the following sense. The size of the list is only quasi-polynomial, but each string in the list is computed in polynomial time.

Theorem 2 (Polynomial-time computable list for increasing the Kolmogorov complexity). *There exists a function f that takes as input $x \in \{0,1\}^*$ and a constant rational number $\delta > 0$ and returns a list of strings of length at most $|x|$ with the following properties:*

1. *The size of the list is bounded by $2^{O(\log|x| \cdot \log(|x|/\delta))}$;*
2. *If $C(x) < |x| - O(\log|x| \cdot \log(|x|/\delta))$, then $(1-\delta)$ fraction of the elements in the list $f(x)$ have a Kolmogorov complexity larger than $C(x)$;*
3. *The function f is computable in polynomial time in the following sense: there is a polynomial time algorithm that takes as input x, i and computes the i-th element in the list $f(x)$.*

Remark 1. *A preliminary version of this paper has appeared in STACS 2017 [1]. In that version, it was claimed that the result in Theorem 1 holds for all strings x with $C(x) < |x|$. The proof had a bug and we can only prove it for strings satisfying $C(x) < |x| - \log\log|x| - O(1)$. The proof of Theorem 2 given here is different from that in [1]. Theorem 2 has better parameters than its analog in the preliminary version.*

Remark 2. *Any procedure that constructs the approximation list can be converted into a probabilistic algorithm that does the same work and picks one random element from the list. The procedure in Theorem 2 can be converted into a polynomial-time probabilistic algorithm, which uses $O(\log|x| \cdot \log(|x|/\delta))$ random bits to pick which element from the list to construct (see item 3 in the statement).*

Vice-versa, a probabilistic algorithm can be converted into a list-approximation algorithm in the obvious way, i.e., by constructing the list that has as elements the outputs of the algorithm for all choices of the random coins.

Thus, a list-approximation algorithm A_1, in which $(1-\delta)$ elements in the list have the desired property, is equivalent to a probabilistic algorithm A_2 that succeeds with probability $1-\delta$. The number of random bits used by A_2 is the logarithm in base two of the size of the list produced by A_1.

1.1. Basic Concepts and Notation

We recall the standard setup for Kolmogorov complexity. We fix an universal Turing machine U. The universality of U means that, for any Turing machine M, there exists a computable "translator" function t, such that, for all strings p, $M(p) = U(t(p))$ and $|t(p)| \leq |p| + O(1)$. For the polynomial-time constructions, we also require that t is polynomial-time computable. If $U(p) = x$, we say that p is a *program* (or *description*) for x. The Kolmogorov complexity of the string x is $C(x) = \min\{|p| \mid p \text{ is a program for } x\}$. If p is a program for x and $|p| \leq C(x) + c$, we say that p is a c-short program for x.

1.2. Related Works

The problem of increasing the Kolmogorov complexity has been studied before by Buhrman, Fortnow, Newman and Vereshchagin [2]. They show that there exists a polynomial-time computable f that takes as input x of length n and returns a list of strings, all having length n, such that, if $C(x) < n$, then there exists y in the list with $C(y) > C(x)$ (this is Theorem 14 in [2]). In the case of complexity conditioned by the string length, they show that it is even possible to compute in polynomial time a list of constant size. That is, $f(x)$ is a list with $O(1)$ strings of length n and, if $C(x \mid n) < n$, then it contains a string y with $C(y \mid n) > C(x \mid n)$ (this is Theorem 11 in [2]). Our results are incomparable with the results in [2]. On one hand, their results work for any input x with complexity less than $|x|$, while, in Theorem 1, we only handle inputs with complexity at most $|x| - \log\log|x| - O(1)$ (and, in Theorem 2, the complexity of the input is required to be even lower). On the other hand, they only guarantee that one string in the output list has higher complexity than x, while we guarantee this property for most strings in the output list and this can be viewed as a probabilistic algorithm with few random bits as explained in Remark 2.

This paper is inspired by recent list-approximation results regarding another problem in the Kolmogorov complexity, namely, the construction of short programs (or descriptions) for strings. Using a Berry paradox argument, it is easy to see that it is impossible to effectively construct a shortest program for x (or, even a, say, $n/2$-short program for x). Remarkably, Bauwens et al. [3] show that effective list approximation for short programs is possible. There is an algorithm that, for some constant c, takes as input x and returns a list with $O(|x|^2)$ strings guaranteed to contain a c-short program for x. They also show a lower bound; The quadratic size of the list is minimal up to constant factors. Bauwens and Zimand [4] consider a more general type of optimal compressor that goes beyond the standard Kolmogorov complexity and, using another type of pseudo-random function called *conductor*, re-obtains the overhead of $O(\log^2 n)$. Theorem 2 directly uses results from the latter, namely, Theorem 3. Theorem 1 uses a novel construction, but some of the ideas are inspired from the papers mentioned above.

2. Technique and Proof Overview

We start by presenting an approach that probably comes to mind first. It does not work for inputs x having a complexity very close to $|x|$, such as in Theorem 1 (for which we use a more complicated argument), but, combined with the results from [4], it yields Theorem 2.

Given that we want to modify a string x so that it becomes more complex, which, in a sense, means more random, a simple idea is to just append a random string z to x. Indeed, if we consider strings z of length c, then $C(xz) > C(x) + c/2$, for most strings z, provided that c is large enough. Let us see why this is true. Let $k = C(x)$ and let z be a string that satisfies the opposite inequality, that is,

$$C(xz) \leq C(x) + c/2, \tag{1}$$

Given a shortest program for xz and a self-delimited representation of the integer c, which is $2\log c$ bits long, we obtain a description of x with at most $k + c/2 + 2\log c$ bits. Note that, in this way, from different z's satisfying (1), we obtain different programs for x that are $(c/2 + 2\log c)$-short. By a theorem of Chaitin [5] (also presented as Lemma 3.4.2 in [6]), for any d, the number of d-short programs for x is bounded by $O(2^d)$. Thus, the number of strings z satisfying (1) is bounded by $2^{c/2+2\log c+O(1)}$. Since, for large c, $2^{c/2+2\log c+O(1)}$ is much smaller than 2^c, it follows that most strings z of length c satisfy the claimed inequality (the opposite of (1)). Therefore, we obtain the following lemma.

Lemma 1. *If we append to a string x, a string z chosen at random in $\{0,1\}^c$, then $C(xz) > C(x) + c/2$ with probability $1 - 2^{-(c/2-2\log c-O(1))}$.*

The problem with appending a random z to x is that this operation not only increases the complexity (which is something we want) but also increases the length (which is something we do not want). The natural way to get around this problem is to first compress x to close to the minimal description length using the probabilistic algorithms from [4] described in the Introduction and then append z. If we know $C(x)$, then the algorithms from [4] compress x to length $C(x) + \Delta(n)$, where n is the length of x and $\Delta(n)$ (called the *overhead*) is $O(\log n)$ (or $\text{poly}(\log n)$ for the polynomial-time algorithm). After appending a random z of length c, we obtain a string of length $C(x) + \Delta(n) + c$ and, for this to be n (so that length is not increased), we need $C(x) \leq n - \Delta(n) - c$. This is the idea that we follow for Theorem 2, with an adjustment caused by the fact that we do not know $C(x)$ but only a bound of it.

However, in this way, we cannot obtain a procedure that works for all x with $C(x) < n - \log\log n - O(1)$, as required in Theorem 1. Our proof for this theorem is based on a different construction. The centerpiece is a type of bipartite graph with a low congestion property. Once we have the graph (in which the two bipartitions are called the set of left nodes and the set of right nodes), we view x as a left node and the list $f(x)$ consists of some of the nodes at distance 2 from x in the graph. (A side remark: Buhrman et al. [2] also use graphs, namely, constant-degree expanders, and they obtain the lists also as the set of neighbors at some given distance.) In our graph, the left side is $L = \{0,1\}^n$, the set of n-bit strings, the right side is $R = \{0,1\}^m$, the set of m-bit strings, and each left node has degree D. The graphs also depend on three parameters, ϵ, Δ and t, and, for our discussion, it is convenient to also use $\delta = \epsilon^{1/2}$ and $s = \delta \cdot \Delta$. The graphs that we need have two properties:

- For every subset B of left nodes of size at most 2^t, the $(1-\delta)$ fraction of nodes in B satisfies the low congestion condition which requires that the $(1-\delta)$ fraction of their right neighbors have at most s neighbors in B. (More formally, for all $B \subseteq L$ with $|B| \leq 2^t$, for all $x \in B$, except at most $\delta|B|$ elements, all neighbors y of x, except at most δD, have $\deg_B(y) \leq s$, where $\deg_B(y)$ is the number of y's neighbors that are in B. We say that such x has the low-congestion property for B.)
- Each right node has at least Δ neighbors.

The graph with the above two properties is constructed using the probabilistic method in Lemma 2.

Let us now see how to use such a graph to increase the Kolmogorov complexity in the list-approximation sense. Let us suppose that we have a graph G with the above properties for the parameters n, δ, Δ, D, s and t.

Claim 1. *There is a procedure that takes as input a string x of length n with complexity $C(x) < t$ and produces a list with $D \cdot \Delta$ strings, all having length n, such that at least a fraction of $(1 - 2\delta)$ of the strings in the list has a complexity larger than $C(x)$.*

Indeed, let x be a string of length n with $C(x) = k < t$. Let us consider the set $B = \{x' \in \{0,1\}^n \mid C(x') \leq k\}$, which we view as a set of left nodes in G. Note that the size of B is bounded by 2^t. A node that does not have the low-congestion property for B is said to be δ-BAD(B). By the first property of G, there are at most $\delta|B|$ elements in B that are δ-BAD(B). It can be shown that x is not δ-BAD(B). The reason is, essentially, that the strings that are δ-BAD(B) can be enumerated and they make up a small fraction of B; therefore, they can be described with less than k bits. Now, to construct the list, we view x as a left node in G and we "go-right-then-go-left". This means that we first "go-right", i.e., we take all the D neighbors of x and, for each such neighbor y, we "go-left", i.e., we take Δ of the y's neighbors and put them in the list. Since x is not δ-BAD(B), $(1 - \delta)D$ of its neighbors have at most $s = \delta \cdot \Delta$ elements in B. Overall, less than $2\delta \cdot D \cdot \Delta$ of the strings in the list can be in B and so at least a fraction of $(1 - 2\delta)$ of the strings in the list has complexity larger than $k = C(x)$. Our claim is proved.

3. Proof of Theorem 2

We use the following definition and results from [4].

Definition 1.

- *A* compressor *$\mathcal{C}$ is a probabilistic function that takes as input a rational number $\epsilon > 0$, a positive integer m and a string x and outputs (with probability 1) a string $\mathcal{C}(\epsilon, m, x)$ of length exactly m.*
- *$\Delta(\epsilon, m, n)$ is a function of ϵ and positive integers m and n, called* overhead.
- *A compressor $\mathcal{C}$ is Δ-optimal for the Kolmogorov complexity, if there exists an algorithm $\mathcal{D}$ (called* decompressor*) such that, for every string x, every rational $\epsilon \geq 2^{-|x|}$ and every $m \geq C(x) + \Delta(\epsilon, m, |x|)$,*

$$Prob[\mathcal{D}(\mathcal{C}(\epsilon, m, x)) = x] \geq 1 - \epsilon.$$

In other words, if we are given a bound m that is at least $C(x)$+overhead, then $\mathcal{C}$ compresses x to a string of length m, from which $\mathcal{D}$ is able to reconstruct x with high probability.

Theorem 3 (Theorem 1.1 in [4])**.** *There exists a compressor $\mathcal{C}$ with overhead $\Delta(\epsilon, m, n) = O(\log m \cdot \log(n/\epsilon))$ that is Δ-optimal for the Kolmogorov complexity. Furthermore, the compressor $\mathcal{C}$ takes as input (ϵ, m, x) and runs in polynomial time in $|x|$, using a random string of length $O(\log m \cdot \log(|x|/\epsilon))$.*

Note: Theorem 1.1 in [4] is more general, but we only need the above version.

Proof of Theorem 2. We follow the plan sketched in Section 2; we compress the input x to a string y with the optimal compressor from Theorem 3 and then append to y a random string z of constant length. We show that, with high probability, yz has the desired properties; it has a complexity larger than $C(x)$ and it is not longer than x. We see below that this randomized algorithm uses $O(\log|x| \cdot \log|x|/\epsilon))$ random bits, which implies the desired list approximation via the observations in Remark 2.

Let the compressor $\mathcal{C}$ and the overhead Δ be the functions from Theorem 3. Let $\epsilon = \delta/2$. We fix n; let us consider a string x of length n such that $C(x) \leq n - 3\Delta(\epsilon, n, n)$. Note that $C(x) \leq n - O(\log n \cdot \log(n/\epsilon))$. Let $m = n - 2\Delta(\epsilon, n, n)$ and $y = \mathcal{C}(\epsilon, m, x)$ (note that y is a random variable because $\mathcal{C}$ is a randomized function). For n sufficiently large,

$$C(x) \leq n - 3\Delta(\epsilon, n, n) \leq m - \Delta(\epsilon, m, n).$$

Let $\mathcal{A}$ be the event by which the decompressor $\mathcal{D}$ reconstructs x from y. By Theorem 3, $\mathcal{A}$ has probability $1-\epsilon$.

We take c a constant large enough such that Equations (2) and (3) below are satisfied. Conditioned by $\mathcal{A}$,

$$C(y) \geq C(x) - c \text{ (because } x \text{ is reconstructed from } y) \tag{2}$$

Let $c' = 2c$. We choose c so that

$$2^{-(c'/2 - 2\log c' - O(1))} < \epsilon, \tag{3}$$

where the $O(1)$ term is the constant from Lemma 1.

We append to y a string z chosen at random in $\{0,1\}^{c'}$. By Lemma 1 and Equation (3), with probability $1-\epsilon$, $C(yz) > C(y) + c'/2 = C(y) + c$. Now, we condition on $\mathcal{A}$ and we obtain that, with probability $1-2\epsilon$,

$$C(yz) > C(y) + c \geq C(x) - c + c = C(x).$$

We take $\delta = 2\epsilon$. Now, let us check the properties of the above algorithm. For every n-bit string x with $C(x) \leq n - 3\Delta(\epsilon, n, n) = n - O(\log|x| \cdot \log|x|/\delta)$, the algorithm takes as input x and δ and outputs, in polynomial time, the string yz that, with probability $1-\delta$, has a complexity larger than the complexity of x. The string yz has length $m + c = n - 2\Delta(\epsilon, n, n) + c \leq n$. The whole randomized procedure uses $O(\log m \cdot \log(n/\epsilon)) = O(\log n \cdot \log(n/\delta))$ random bits for compression with $\mathcal{C}$ and $c' = O(1)$ random bits for z. The list approximation is obtained from the probabilistic algorithm in the obvious way, i.e., by including in the list one element for each choice of the random string (see Remark 2). The theorem is proved. □

4. Proof of Theorem 1

We split the proof in three parts. In Section 4.1, we introduce *balanced graphs*; in Section 4.2, we show how to increase the Kolmogorov complexity in the list approximation sense using balanced graphs and, in Section 4.3, we use the probabilistic method to obtain the balanced graph with the parameters needed for Theorem 1.

4.1. Balanced Graphs

Here, we formally define the type of graphs that we need. We work with families of bipartite graphs $G_n = (L \cup R, E \subseteq L \times R)$, indexed by n, which have the following structure:

1. The vertices are labeled with binary strings, $L = \{0,1\}^n$ and $R = \{0,1\}^n$, where we view L as the set of left nodes and R as the set of right nodes.
2. All the left nodes have the same degree D; $D = 2^d$ is a power of two and the edges outgoing from a left node x are labeled with binary strings of length d.
3. We allow multiple edges between two nodes to exist. For a node x, we write $N(x)$ for the *multiset* of x's neighbors, each element being taken with the multiplicity equal to the number of edges from x landing into it.

A bipartite graph of this type can be viewed as a function $\mathrm{EXT} : \{0,1\}^{\mathrm{n}} \times \{0,1\}^{\mathrm{d}} \to \{0,1\}^{\mathrm{n}}$, where $\mathrm{EXT}(\mathrm{x},\mathrm{y}) = \mathrm{z}$ if there is an edge between x and z labeled y. We want EXT to yield a (k, ϵ) randomness extractor whenever we consider the modified function $\mathrm{EXT}_{\mathrm{k}}$, which takes as input (x, y) and returns $\mathrm{EXT}(\mathrm{x},\mathrm{y})$, from which we keep only the first k bits. (Note: A randomness extractor is a type of function that plays a central role in the theory of pseudo-randomness. All we need here is that it satisfies Equation (4).)

From the function $\mathrm{EXT}_{\mathrm{k}}$, we go back to the graph representation and we obtain the "prefix" bipartite graph $G_{n,k} = (L = \{0,1\}^n, R_k = \{0,1\}^k, E_k \subseteq L \times R_k)$, where, in $G_{n,k}$, we merge the right nodes of G_n that have the same prefix of length k. The left degrees in the prefix graph do not change. However, the right degrees may change and, as k becomes smaller, the right degrees typically become larger due to merging.

The requirement is that, for every subset $B \subseteq L$ of size $|B| \geq 2^k$, for every $A \subseteq R_k$,

$$\left| \frac{|E_k(B, A)|}{|B| \times D} - \frac{|A|}{|R_k|} \right| \leq \epsilon, \tag{4}$$

where $E_k(B, A)$ is the set of edges between B and A in $G_{n,k}$. (Note: This means that $G_{n,k}$ is a (k, ϵ) randomness extractor.)

We also want to have the guarantee that each right node in $G_{n,t}$ has degree at least Δ, where Δ and t are parameters.

Accordingly, we have the following definition.

Definition 2. *A graph $G_n = (L, R, E \subseteq L \times R)$ as above is (ϵ, Δ, t)-balanced if the following requirements hold:*

1. *For every $k \in \{1, \ldots, n\}$, let $G_{n,k}$ be the graph corresponding to $\mathrm{EXT_k}$ described above. We require that, for every $k \in \{1, \ldots, n\}$, $G_{n,k}$ is a (k, ϵ) extractor, i.e., $G_{n,k}$ has the property in Equation (4).*
2. *In the graph $G_{n,t}$, every right node with non-zero degree has degree at least Δ.*

In our application, we need balanced graphs in which the neighbors of a given node can be found effectively. As usual, we consider families of graphs $(G_n)_{n \geq 1}$ and we say that such a family is *computable* if there is an algorithm that takes as input (x, y), views x as a left node in $G_{|x|}$, views y as the label of an edge outgoing from x and outputs z, where z is the right node where the edge y lands in $G_{|x|}$.

The following lemma provides the balanced graphs that we need as explained in the proof overview in Section 2.

Lemma 2. *For every rational $\epsilon > 0$, there exist some constant c and a computable family of graphs $(G_n)_{n \geq 1}$, where each $G_n = (L = \{0,1\}^n, R = \{0,1\}^n, E \subseteq L \times R)$ is (ϵ, Δ, t)-balanced graph, with left degree $D = 2^d$ for $d = \lceil \log(2n/\epsilon^2) \rceil$, $\Delta = 2(1/\epsilon)^{3/2} D$ and $t = n - \log \log n - c$.*

The proof of Lemma 2 is by the standard probabilistic method and is presented in Section 4.3.

4.2. From Balanced Graphs to Increasing the Kolmogorov Complexity in the List-Approximation Sense

The following lemma shows a generic transformation of a balanced graph into a function that takes as input x and produces a list so that most of its elements have a complexity larger than $C(x)$.

Lemma 3. *Let us suppose that, for every $\delta > 0$, there are $t = t(n)$ and a computable family of graphs $(G_n)_{n \geq 1}$, where each $G_n = (L_n = \{0,1\}^n, R_n = \{0,1\}^n, E_n \subseteq L_n \times R_n)$ is (δ^2, Δ, t)-balanced graph, with $\Delta = 2(1/\delta^3) \cdot D$, where D is the left degree.*

Then, there exists a computable function f that takes as input a string x and a rational number $\delta > 0$ and returns a list containing strings of length $|x|$; additionally, the following are true:

1. *The size of the list is $O((1/\delta)^3 D^2)$;*
2. *If $C(x) \leq t$, then $(1 - O(\delta))$ of the elements in the list have a complexity larger than $C(x)$.*

(The constants hidden in $O(\cdot)$ do not depend on δ.)

Proof. The following arguments are valid if δ is smaller than some small positive constant. We assume that δ satisfies this condition and also that it is a power of $1/2$. This can be performed because scaling down δ by a constant factor only changes the constants in the

$O(\cdot)$ in the statement. Let $\epsilon = \delta^2$. We explain how to compute the list $f(x)$, with the property stipulated in the theorem's statement.

We take G_n to be the (ϵ, Δ, t)-balanced graph with left nodes of length n promised by the hypothesis. Let $G_{n,t}$ be the "prefix" graph obtained from G_n by cutting the last $n - t$ bits in the labels of right nodes (thus preserving the prefix of length t in the labels).

The list $f(x)$ is computed in two steps:

1. First, we view x as a left node in $G_{n,t}$ and take $N(x)$, the multiset of all neighbors of x in $G_{n,t}$.
2. Secondly, for each p in $N(x)$, we take A_p to be a set of Δ neighbors of p in $G_{n,t}$ (e.g., the first Δ ones in some canonical order). We set $f(x) = \bigcup_{p \in N(x)} A_p$ (if p appears n_p times in $N(x)$, we also take A_p in the union n_p times; note that $f(x)$ is a multiset).

Note that all the elements in the list have length n and the size of the list is $|f(x)| = \Delta \cdot D = 2(1/\delta)^3 D^2$.

Let x be a binary string of length n, with complexity $C(x) = k$. We assume that $k \leq t$. The rest of the proof is dedicated to showing that the list $f(x)$ satisfies the second item in the statement. Let

$$B_{n,k} = \{x' \in \{0,1\}^n \mid C(x') \leq k\},$$

and let $S_{n,k} = \lfloor \log |B_{n,k}| \rfloor$. Thus, $2^{S_{n,k}} \leq |B_{n,k}| < 2^{S_{n,k}+1}$. Later, we use the fact that

$$S_{n,k} \leq k \leq t. \quad (5)$$

We consider the graph $G_{n,S_{n,k}}$, which is obtained, as explained above, from G_n by taking the prefixes of the right nodes of length $S_{n,k}$. To simplify notation, we use G instead of $G_{n,S_{n,k}}$. The set of left nodes in G is $L = \{0,1\}^n$ and the set of right nodes in G is $R = \{0,1\}^m$, for $m = S_{n,k}$.

We view $B_{n,k}$ as a subset of the left nodes in G. Let us introduce some helpful terminology. In the following, all the graph concepts (left node, right node, edge and neighbor) refer to the graph G. We say that a right node z in G is $(1/\epsilon)$-light if it has at most $(1/\epsilon) \cdot \frac{|B_{n,k}| \cdot D}{|R|}$ neighbors in $B_{n,k}$. A node that is not $(1/\epsilon)$-light is said to be $(1/\epsilon)$-heavy. Note that

$$(1/\epsilon) \cdot \frac{|B_{n,k}| \cdot D}{|R|} \leq (1/\epsilon) \frac{2^{S_{n,k}+1} \cdot D}{2^{S_{n,k}}} = \delta\Delta,$$

thus, a $(1/\epsilon)$-light node has at most $\delta\Delta$ neighbors in $B_{n,k}$.

We also say that a left node in $B_{n,k}$ is δ-BAD with respect to $B_{n,k}$ if at least a δ fraction of the D edges outgoing from it lands in the right neighbors that are $(1/\epsilon)$-heavy. Let δ-BAD$(B_{n,k})$ be the set of nodes that are δ-BAD with respect to $B_{n,k}$.

We show the following claim.

Claim 2. *At most a 2δ fraction of the nodes in $B_{n,k}$ is δ-BAD with respect to $B_{n,k}$.*

(In other words, for every x' in $B_{n,k}$ except at most a 2δ fraction, at least a $(1-\delta)$ fraction of the edges going out from x' in G lands in the right nodes that have at most Δ' neighbors with complexity at most k).

We defer for later the proof of Claim 2 and continue the proof of the theorem.

For any positive integer k, let

$$B_k = \{x' \mid C(x') \leq k \text{ and } k \leq t(|x'|)\}.$$

Let $I_k = \{n \mid k \leq t(n)\}$. Note that $|B_k| = \sum_{n \in I_k} |B_{n,k}|$. Let $x' \in B_k$ and let $n' = |x'|$. We say that x' is δ-BAD with respect to B_k if, in $G_{n'}$, x' is δ-BAD with respect to $B_{n',k}$. We denote by δ-BAD(B_k) the set of nodes that are δ-BAD with respect to B_k. We upper bound the size of δ-BAD(B_k) as follows:

$$\begin{aligned} |\delta\text{-BAD}(B_k)| &= \textstyle\sum_{n' \in I_k} |\delta\text{-BAD}(B_{n',k})| \\ &\leq \textstyle\sum_{n' \in I_k} 2\delta \cdot |B_{n',k}| \qquad \text{(by Claim 2)} \\ &= 2\delta \textstyle\sum_{n \in I_k} |B_{n',k}| \\ &= 2\delta |B_k| \\ &\leq 2\delta \cdot 2^{k+1}. \end{aligned}$$

Note that the set δ-BAD(B_k) can be enumerated given k and δ. Therefore, a node x' that is δ-BAD with respect to B_k can be described by k, δ and its ordinal number in the enumeration of the set δ-BAD(B_k). We write the ordinal number on exactly $k+2-\log(1/\delta)$ bits and δ in a self-delimited way on $2\log\log(1/\delta)$ bits (recall that $1/\delta$ is a power of 2), so that k can be inferred from the ordinal number and δ. It follows that, if x' is δ-BAD with respect to B_k, then, provided $1/\delta$ is sufficiently large,

$$C(x') \leq k + 2 - \log(1/\delta) + 2\log\log(1/\delta) + O(1) < k. \tag{6}$$

Now, we recall our string $x \in \{0,1\}^n$, which has complexity $C(x) = k$. The inequality (6) implies that x cannot be δ-BAD with respect to B_k, which means that $(1-\delta)$ of the edges going out from x land in neighbors in G having at most $\delta\Delta$ neighbors in B_k. The same is true if we replace G by $G_{n,t}$, because, by the inequality (5), the right nodes in G are prefixes of the right nodes in $G_{n,t}$.

Now, let us suppose that we pick at random a neighbor p of x in $G_{n,t}$ and then find a set A_p of Δ neighbors of p in $G_{n,t}$. Then, with probability $1-\delta$, only a fraction of δ of the elements of A_p can be in B_k. Let us recall that we have defined the list $f(x)$ to be

$$f(x) = \bigcup_{p \text{ neighbor of } x \text{ in } G_{n,t}} A_p.$$

It follows that at least a $(1-\delta)^2 > (1-2\delta)$ fraction of the elements in $f(x)$ has complexity larger than $C(x)$. This ends the proof. □

We now prove Claim 2.

Proof of Claim 2. Let A be the set of right nodes that are $(1/\epsilon)$-heavy. Then,

$$|A| \leq \epsilon |R|.$$

Indeed, the number of edges between $B_{n,k}$ and A is at least $|A| \cdot (1/\epsilon) \cdot \frac{|B_{n,k}| \cdot D}{|R|}$ (by the definition of $(1/\epsilon)$-heavy), but, at the same time, the total number of edges between $B_{n,k}$ and R is $|B_{n,k}| \cdot D$ (because each left node has degree D).

Next, we show that

$$|\delta\text{-BAD}(B_{n.k})| \leq 2\delta |B_{n,k}|. \tag{7}$$

For this, note that G is a $(S_{n,k}, \epsilon)$ randomness extractor and $B_{n,k}$ has size at least $2^{S_{n,k}}$. Therefore, by the property (4) of extractors,

$$\frac{|E(B_{n,k}, A)|}{|B_{n,k}| \cdot D} \leq \frac{|A|}{|R|} + \epsilon \leq 2\epsilon.$$

On the other hand, the number of edges linking $B_{n,k}$ and A is at least the number of edges linking δ-BAD$(B_{n,k})$ and A; this number is at least $|\delta\text{-BAD}(B_{n,k})| \cdot \delta D$. Thus,

$$|E(B_{n,k}, A)| \geq |\delta\text{-BAD}(B_{n,k})| \cdot \delta D.$$

Combining the last two inequalities, we obtain

$$\frac{|\delta\text{-BAD}(B_{n,k})|}{|B_{n,k}|} \leq 2\epsilon \cdot \frac{1}{\delta} = 2\delta.$$

This ends the proofs of Claim 2, which is the last piece that we needed for the proof of Lemma 3. □

Theorem 1 is obtained by plugging, into the above lemma, the balanced graphs from Lemma 2 with parameter $\epsilon = \delta^2$.

4.3. *Construction of Balanced Graphs: Proof of Lemma 2*

We use the probabilistic method. We consider a random function $\mathrm{EXT} : \{0,1\}^{\mathrm{n}} \times \{0,1\}^{\mathrm{d}} \to \{0,1\}^{\mathrm{n}}$ for $d = \lceil \log(2n/\epsilon^2) \rceil$. We show the following two claims, which imply that a random function has the desired properties with positive probability. Since the properties can be checked effectively, we can find a graph by exhaustive search. We use the notation from Definition 2 and from the paragraph preceding it.

Claim 3. *For sufficiently large n, with probability $\geq 3/4$, it holds that, for every $k \in \{1, \ldots, n\}$, in the bipartite graph $G_{n,k} = \{L, R_k, E_k \subseteq L \times R_k\}$, every $B \subseteq L = \{0,1\}^n$ of size $|B| \geq 2^k$ and every $A \subseteq R_k = \{0,1\}^k$ satisfies*

$$\left| \frac{|E_k(B,A)|}{|B| \times D} - \frac{|A|}{|R_k|} \right| \leq \epsilon. \tag{8}$$

Claim 4. *For some constant c and every sufficiently large positive integer n, with probability $\geq 3/4$, every right node in the graph $G_{n,n-\log\log n - c}$ has degree at least Δ.*

Proof of Claim 3. First, we fix $k \in \{1, \ldots, n\}$ and let $K = 2^k$ and $N = 2^n$. Let us consider $B \subseteq \{0,1\}^n$ of size $|B| \geq K$ and $A \subseteq R_k$. For a fixed $x \in B$ and $y \in \{0,1\}^d$, the probability that $\mathrm{EXT_k(x,y)}$ is in A is $|A|/|R_k|$. By the Chernoff bounds,

$$\mathrm{Prob}\left[\left| \frac{|E_k(B,A)|}{|B| \times D} - \frac{|A|}{|R_k|} \right| > \epsilon \right] \leq 2^{-\Omega(K \cdot D \cdot \epsilon^2)}.$$

The probability that relation (8) fails for a fixed k, some $B \subseteq \{0,1\}^k$ of size $|B| \geq K$ and some $A \subseteq R_k$ is bounded by $2^K \cdot \binom{N}{K} \cdot 2^{-\Omega(K \cdot D \cdot \epsilon^2)}$, because A can be chosen in 2^K ways; further, we can consider that B has size exactly K and that there are $\binom{N}{K}$ possible choices of such B's. Since $D \geq 2n/\epsilon^2$, the above probability is much less than $(1/4)2^{-k}$. Therefore, the probability that relation (8) fails for some $k \in \{1, \ldots, n\}$, some B and some A is less than $1/4$. □

Proof of Claim 4. We use a "coupon collector" argument. We consider the graph $G_{n,n-\log\log n - c}$ for some constant c to be fixed later. This graph is obtained from the above function EXT as explained in Definition 2. The graph $G_{n,n-\log\log n - c}$ is a bipartite graph with left side $L = \{0,1\}^n$, right side $R' = \{0,1\}^{n-\log\log n - c}$ and each left node has degree $D = 2^d$. We show that, with probability $\geq 3/4$, every right node in $G_{n,n-\log\log n - c}$ has degree at least Δ. The random process consists of drawing, for each $x \in L$ and edge $y \in \{0,1\}^d$, a random element from R'. Thus, we draw at random ND times, with replacement, from a set with $|R'|$ "coupons". Newman and Shepp [7] have shown that, to obtain at least h times each coupon from a set of p coupons, the expected number of draws is $p \log p + (h-1) p \log\log p + o(p)$. By Markov's inequality, if the number of draws is 4 times the expected value, we collect each coupon p times with probability $3/4$. In our case, we have $p = 2^{n-\log\log n - c}$ and $h = \Delta$; it can be checked readily that, for an appropriate choice of the constant c, $4(p \log p + (h-1) p \log\log p + o(p)) < ND$, provided n is large enough. □

Funding: The author has been supported in part by the National Science Foundation through grant CCF 1811729.

Institutional Review Board Statement: Not applicable.

Informed Consent Statement: Not applicable.

Data Availability Statement: Not applicable.

Acknowledgments: The author is grateful to Bruno Bauwens for his insightful observations and to Nikolay Vereshchagin for pointing out an error in an earlier version. The author thanks the anonymous referees for their useful suggestions.

Conflicts of Interest: The author declares no conflict of interest..

References

1. Zimand, M. List Approximation for Increasing Kolmogorov Complexity. In Proceedings of the 34th Symposium on Theoretical Aspects of Computer Science, STACS 2017, Hannover, Germany, 8–11 March 2017; Vollmer, H., Vallée, B., Eds.; Schloss Dagstuhl-Leibniz-Zentrum für Informatik: Dagstuhl, Germany, 2017; Leibniz International Proceedings in Informatics (LIPIcs); Volume 66, pp. 58:1–58:12. [CrossRef]
2. Buhrman, H.; Fortnow, L.; Newman, I.; Vereshchagin, N. Increasing Kolmogorov complexity. In Proceedings of the 22nd Annual Symposium on Theoretical Aspects of Computer Science, Stuttgart, Germany, 24–26 February 2005; Lecture Notes in Computer Science #3404; Springer: Berlin, Germany, 2005; pp. 412–421.
3. Bauwens, B.; Makhlin, A.; Vereshchagin, N.; Zimand, M. Short lists with short programs in short time. In Proceedings of the 28th IEEE Conference on Computational Complexity, Stanford, CA, USA, 5–7 June 2013.
4. Bauwens, B.; Zimand, M. Universal almost optimal compression and Slepian-Wolf coding in probabilistic polynomial time. *arXiv* **2019**, arXiv:1911.04268.
5. Chaitin, G.J. Information-Theoretic Characterizations of Recursive Infinite Strings. *Theor. Comput. Sci.* **1976**, *2*, 45–48. [CrossRef]
6. Downey, R.; Hirschfeldt, D. *Algorithmic Randomness and Complexity*; Springer: New York, NY, USA, 2010.
7. Newman, D.; Shepp, L. The Double Dixie Cup Problem. *Am. Math. Mon.* **1960**, *67*, 58–61. [CrossRef]

axioms

MDPI

Article

Two Extensions of Cover Automata

Cezar Câmpeanu

School of Mathematical and Computational Sciences, University of Prince Edward Island, 550 University Ave, Charlottetown, PE C1A 4P3, Canada; ccampeanu@upei.ca

Abstract: Deterministic Finite Cover Automata (DFCA) are compact representations of finite languages. Deterministic Finite Automata with "do not care" symbols and Multiple Entry Deterministic Finite Automata are both compact representations of regular languages. This paper studies the benefits of combining these representations to get even more compact representations of finite languages. DFCAs are extended by accepting either "do not care" symbols or considering multiple entry DFCAs. We study for each of the two models the existence of the minimization or simplification algorithms and their computational complexity, the state complexity of these representations compared with other representations of the same language, and the bounds for state complexity in case we perform a representation transformation. Minimization for both models proves to be NP-hard. A method is presented to transform minimization algorithms for deterministic automata into simplification algorithms applicable to these extended models. DFCAs with "do not care" symbols prove to have comparable state complexity as Nondeterministic Finite Cover Automata. Furthermore, for multiple entry DFCAs, we can have a tight estimate of the state complexity of the transformation into equivalent DFCA.

Keywords: finite languages; deterministic finite cover automata; multiple entry automata; automata with "do not care" symbols; similarity relations

Citation: Câmpeanu, C. Two Extensions of Cover Automata. *Axioms* **2021**, *10*, 338. https://doi.org/10.3390/axioms10040338

Academic Editor: Oscar Humberto Montiel Ross

Received: 30 September 2021
Accepted: 7 December 2021
Published: 10 December 2021

1. Introduction

The concept of Cover Automata was first presented at a conference paper of Câmpeanu et al. at the Workshop on Implementations and Applications of Automata (WIAA) in Rouen (1999) [1,2] when the authors introduced a formal definition of a Deterministic Finite Cover Automaton (DFCA) and a minimization algorithm. A cover language for a language L is a superset L' of L. If L is a finite non-empty language, then the length of the longest word in L exists, and we can denote it with a natural number, l. A DFCA for a finite language L is a deterministic finite automaton (DFA) accepting a cover language for L, such that the accepted words that are not in L have their length greater than l.

During the last two decades, several papers used DCFAs for compact representation of finite languages. Other efficient minimization algorithms were also published, for example [3–8]. The concept of DFCA was also generalized to the nondeterministic version in a paper presented at AFL 2014 in Szeged by Câmpeanu [9], followed by the journal version [10].

Using nondeterminism, we can reduce the size of the automata recognizing some languages, but minimizing such automata is known to be PSPACE-complete. Therefore, several other intermediate representations of languages that maintain deterministic transitions were proposed. That is why it is a must to study these extensions in case they are applied to cover automata, which we are doing in Section 2. The first extension considered here is to enhance DFCAs with *"do not care"* symbols, thus obtaining finite cover automata with "do not care" symbols, denoted by $\diamond$-DFCAs, in other words, finite cover automata accepting partial words. Fischer and Paterson introduced partial words in [11] in 1974, and the authors in [12,13] prove that the minimization of finite automata with "do not care" symbols is NP-hard. As emphasized by Professor Solomon Marcus in [14], many researches from other areas studied the same concept, but in a different theoretical setup with different notations. An example of such a paper related to partial words is [15], where the authors show strong connections between graph problems and pattern matching with some of the symbols in the patterns not known. In that paper, the "do not care" symbol denoted here by $\diamond$ is denoted by ϕ. We will prefer the $\diamond$ notation because most references use this notation and using a symbol that is not part of any alphabet is easier to identify. Holzer et al. in [16,17] prove that "almost all problems related to partial word automata, such as equivalence and universality, are already PSPACE-complete". Some of their proofs link non-deterministic automata problems with graph theory problems in a similar fashion as it is done in [15]. As such, because the minimization of $\diamond$-DFAs is hard, only the simplification algorithms were developed for these types of finite machines, and an example is presented in [13]. A simplification algorithm will eventually produce an equivalent automaton with less states than the input automaton, but it is not guaranteed to be minimal. In Section 3, we show that the same difficulties found for NFCA's simplification are also present for $\diamond$-DFCAs, even though $\diamond$-DFCAs can be considered a particular simpler class of NFCAs. We show a simplification algorithm for $\diamond$-DFCAs that has a better time complexity than the one presented for $\diamond$-DFAs in [13].

In [12], the authors give an example of automaton having limited nondeterminism—there is only one transition with degree 2 for the same letter—which is hard to minimize. The same argument can be used to prove that finding the minimal finite cover automaton with *"do not care"* symbols is also a hard problem. We already know that NFCA minimization is NP-hard, and details of why the previous proofs work as well for $\diamond$-DFCAs are presented in Section 3. In the same paper [12], the example of an automaton that is hard to minimize accepts a finite language having all the accepted words of length at most 3. This example is used to show that minimizing multiple entry deterministic automata (MEFA) (When the number k of entries is known, we use the term k-DFA instead of MEFA) is hard. In Section 4, we show that the method with exactly the same construction will also work for Multiple Entry Finite Cover Automata (MEFCA), or k-entry DFCAs (k-DFCA), adding the results on k-entry FA to the previous ones obtained in [18–21]. In Section 4, we show that for binary alphabets, by transforming a k-DFCA into a minimal DFCA we can reach the upper bound for NFA to DFCA transformation. Moreover, we show that the general bound is reached for the state complexity of this transformation. Section 5 includes future work and a list of open problems, and the conclusions are drawn in Section 6.

2. Cover Automata Extensions

2.1. Notations

The number of elements of a set T is $\#T$, an alphabet is usually denoted by Σ, and the set of words over Σ is $\Sigma^\star$. The length of a word $w \in \Sigma^\star$ is the number of letters of w, and it is denoted by $|w|$. Thus, if $w = w_1 w_2 \cdots w_k$, where $w_i \in \Sigma$, for all $1 \leq i \leq k$, then $|w| = k$. In particular, when $k = 0$, we have a word with no letters, denoted by ε, and $|\varepsilon| = 0$. We also use the following notations: $\Sigma^{=l} = \{w \in \Sigma^\star \mid |w| = l\}$, $\Sigma^{\leq l} = \{w \in \Sigma^\star \mid |w| \leq l\}$, $\Sigma^{>l} = \{w \in \Sigma^\star \mid |w| > l\}$, $\Sigma^{<l} = \{w \in \Sigma^\star \mid |w| < l\}$, and $\Sigma^+ = \{w \in \Sigma^\star \mid |w| \neq \varepsilon\} = \bigcup_{i>0} \Sigma^i$.

A Deterministic Finite Automaton (DFA) is a quintuple $A = (Q, \Sigma, \delta, q_0, F)$, where Q is a finite non-empty set, the set of states, Σ is the alphabet, $q_0 \in Q$ is the initial state, $F \subseteq Q$ is the set of final states, and $\delta : Q \times \Sigma \longrightarrow Q$ is the transition function. In case the transition function δ is a partial function, denoted as $\delta : Q \times \Sigma \stackrel{\circ}{\longrightarrow} Q$, we have a partial DFA. In case δ is defined for all values of $s \in Q$ and $a \in \Sigma$, the DFA is complete. If we do not emphasize that the DFA is partial, then we understand that the DFA is complete. The transition function δ can be extended in a natural way to $Q \times \Sigma^*$ as follows: $\overline{\delta}(q, \varepsilon) = q$, $\overline{\delta}(q, wa) = \delta(\overline{\delta}(q, w), a)$. For the rest of the paper we denote the extension $\overline{\delta}$, by δ. If the transition function $\delta : Q \times \Sigma \stackrel{\circ}{\longrightarrow} Q$ is a partial function, then the automaton is incomplete; otherwise, it is a complete one. A Nondeterministic Finite Automaton (NFA) is a quintuple

$A = (Q, \Sigma, \delta, Q_0, F)$, where all the elements are the same as for a DFA except, $Q_0 \subseteq Q$ is the set of initial states and the transition function δ, which is now defined as $\delta : Q \times \Sigma \longrightarrow 2^Q$. In case of an NFA, for the transition function we have that for $w \in \Sigma^\star$, $\delta(Q_0, w) \subseteq Q$, and $\overline{\delta}(q, wa) = \bigcup_{s \in \overline{\delta}(q,w)} \delta(s, a)$. In what follows, we will use the NFA's with only one initial state as it is defined in [22]. For any NFA $A = (Q, \Sigma, \delta, Q_0, F)$ there is an equivalent NFA $A = (Q \cup \{q_0\}, \Sigma, \delta', q_0, F)$, where $q_0 \notin Q$, $\delta'(s, a) = \delta(s, a)$, for all $s \in Q$ and $a \in \Sigma$, and $\delta'(q_0, a) = \bigcup_{s \in Q_0} \delta(s, a)$. Using the form for NFA with only one initial state, or initially connected NFAs, simplify most of the definitions and results and the state complexity will differ from the general case by just one state.

For multiple entry automata, we have a quintuple $A = (Q, \Sigma, \delta, Q_0, F)$, where $Q_0 \subseteq Q$. In some cases [23], all the states are considered initial states, thus $Q_0 = Q$, while in most other cases, we consider k-entry DFA so the transition function δ is deterministic and $\#Q_0 = k$ [19].

A state s in a finite automaton A is reachable if there is a word $w \in \Sigma^\star$ such that $\delta(q_0, w) = s$. In case of a k-entry DFA or an NFA, the state q_0 must be one of the initial states. A state s is useful if there exists $w \in \Sigma^\star$ such that $\delta(s, w) \cap F \neq \emptyset$. In case of a deterministic δ, we have that $\delta(s, w) \in F$. A sink state or a dead state is a reachable state with all its transitions being self-loops. All states that are not reachable and not useful can be eliminated without changing the language accepted by the automaton. A deterministic automaton with all states reachable and useful, except one sink state, is called a reduced automaton. In the case of nondeterministic automata, an automaton is considered reduced if all its states are both reachable and useful. In what follows, all automata are reduced automata, so they do not have unreachable or unuseful states.

For an alphabet Σ, we can consider a new symbol $\diamond$, called "do not care symbol", which can replace any letter of Σ. Thus, a word w over the alphabet $\Sigma_\diamond = \Sigma \cup \{\diamond\}$, will be a partial word if $|w|_\diamond > 0$. We say that the word $u \in \Sigma_\diamond^\star$ is weaker than $v \in \Sigma_\diamond^\star$, denoted $u \precsim v$, if $|u| = |v|$ and for all positions i, $1 \leq i \leq |u|$, if $u_i \in \Sigma$ then $u_i = v_i$.

Let $L_\diamond$ be a regular language over the alphabet $\Sigma \cup \{\diamond\}$, with $\sigma : \Sigma_\diamond \longrightarrow 2^\Sigma$ a substitution such that $\sigma(a) = \{a\}$ for all $a \in \Sigma$, and $\sigma(\diamond) \subseteq \Sigma$. The regular language $L_\diamond \subseteq \Sigma_\diamond^\star$ is recognized by $\diamond$-DFA, $A = (Q, \Sigma_\diamond, \delta, q_0, F)$, if $L_\diamond = L(A)$. Accordingly, a $\diamond$-DFA $A_\sigma = (Q, \Sigma_\diamond, \delta, q_0, F)$ associated with some substitution σ is defined as a DFA that recognizes a partial language $L_\diamond$, and it is also associated with the total language $\sigma(L_\diamond)$.

A cover automaton for a finite language L is a DFA recognizing a cover language L' such that $L = L' \cap \Sigma^{\leq l}$, for l being the length of the longest word in L. An l-NFCA A is a cover automaton for the language $L(A) \cap \Sigma^{\leq l}$, [10,24]. Any DFA A accepting a finite language is a DFCA for $L(A)$ with $l = \max\{|w| \mid w \in L(A)\}$.

Two words, x and y, are similar with respect to the finite language L, written $x \sim_L y$, if for every $w \in \Sigma^{\leq l - \max\{|x|,|y|\}}$, $xw \in L$, whenever $yw \in L$. In this definition, l is the length of the longest words in L. The similarity relation on words is not an equivalence relation, as it is only reflexive, symmetric, and semi-transitive.

If A is a DFCA for the finite language L, we can also define the level of a state as the length of the shortest path from the initial state to that state, that is $level_A(p) = \min\{|w| \mid \delta(q_0, w) = p\}$. In case of multiple entry DFCAs, a state will have k levels, i.e., $level_{A,i}(p) = \min\{|w| \mid \delta(q_{0,i}, w) = p\}$, for all $1 \leq i \leq k$, and $level_A(p) = (level_{A,1}(p), level_{A,2}(p), \ldots, level_{A,k}(p))$, where $Q_0 = \{q_{0,1}, \ldots, q_{0,k}\}$.

The following definition is in [10] (Definition 2):

Definition 1. *In a NFCA $A = (Q, \Sigma, \delta, q_0, F)$, two states $p, q \in Q$ are similar, written $s \sim_A q$, if $\delta(p, w) \cap F \neq \emptyset$ if $\delta(q, w) \cap F \neq \emptyset$, for all $w \in \Sigma^{\leq l - \max\{level(p), level(q)\}}$.*

In case the NFCA A is understood, we may omit the subscript A, i.e., we write $p \sim q$ instead of $p \sim_A q$, also we can write $level(p)$ instead of $level_A(p)$.

We consider only non-trivial NFCAs for L, i.e., NFCAs such that $level(p) \leq l$ for all states p, and with all the states useful and reachable.

We define deterministic and nondeterministic state complexity of a language as:

$$sc(L) = \min\{\#Q \mid A = (Q, \Sigma, \delta, q_0, F), \text{ is deterministic, complete, and } L = L(A)\}$$

and

$$nsc(L) = \min\{\#Q \mid A = (Q, \Sigma, \delta, q_0, F), \text{is non-deterministic and } L = L(A)\}.$$

In case of a finite language L, we can also define the cover complexity variants:

$$csc(L) = \min\{\#Q \mid A = (Q, \Sigma, \delta, q_0, F), \text{ deterministic, complete, and } L = L(A) \cap \Sigma^{\leq l}\}$$

and

$$ncsc(L) = \min\{\#Q \mid A = (Q, \Sigma, \delta, q_0, F), \text{ non-deterministic, and } L = L(A) \cap \Sigma^{\leq l}\}.$$

We have that $ncsc(L) \leq nsc(L) \leq sc(L)$, and $ncsc(L) \leq csc(L) \leq sc(L)$.

For an automaton A, we say that it is minimal if the number of states of A is equal to its corresponding complexity; therefore, we can have minimal DFAs, NFAs, DFCAs, NFCAs, $\diamond$-DFAs, MEFAs, and MEFCAs. An algorithm which takes an automaton of one of the above types as input and produces a minimal automaton of the same type as output is called a minimization algorithm. In some cases, minimization algorithms are exponential. Therefore, it is worth designing algorithms that will reduce the number of states, but they may not produce a minimal automaton. In that case, we have simplification algorithms that may reduce the number of states of the automaton used as input and produce an equivalent one with possibly fewer states. Simplification algorithms are preferred for cases when their computational complexity is significantly lower than the complexity of a minimization algorithm.

For either minimization or simplification algorithms, the method that is used the most is to merge two or more states into one state in such a way that the recognized language does not change. By merging state p into state q we redirect all incoming transitions to state p to incoming transitions to state q. For outgoing transitions in case of deterministic automata, outgoing transitions from p are lost, but outgoing transitions from q are preserved. In case of nondeterministic automata merging can be done in many different ways. For example, the following definition is in [10] (Definition 3):

Definition 2. *Let $A = (Q, \Sigma, \delta, q_0, F)$ be an NFCA for the finite language L.*

1. *We say that the state q is* weakly mergeable *in state p if the automaton $A' = (Q', \Sigma, \delta', q_0, F')$, where $Q' = Q - \{q\}$, $F' = F \cap Q'$, and*

$$\delta'(s,a) = \begin{cases} \delta(s,a), & \text{if } \delta(s,a) \subseteq Q' \text{ and } s \neq p, \\ (\delta(s,a) \setminus \{q\}) \cup \{p\}, & \text{if } q \in \delta(s,a) \text{ and } s \neq p, \\ (\delta(s,a) \cup \delta(q,a)) \setminus \{q\}, & \text{if } s = p \end{cases}$$

is also an NFCA for L. In this case, we write $p \precapprox q$.

2. *We say that the state q is* strongly mergeable *in state p, if the automaton $A' = (Q', \Sigma, \delta', q_0, F')$, where $Q' = Q - \{q\}$, $F' = F \cap Q'$, and*

$$\delta'(s,a) = \begin{cases} \delta(s,a), & \text{if } \delta(s,a) \subseteq Q' \\ (\delta(s,a) \setminus \{q\}) \cup \{p\}, & \text{if } q \in \delta(s,a), \end{cases}$$

is also an NFCA for L. In this case, we write $p \precsim q$.

By Theorem 3 of [10], if two states are similar, they are also strongly mergeable; therefore, we can reduce the size of that automaton.

Next, we analyze two possible extensions of cover automata. One of them is to allow "do not care" symbols, while the other is to add multiple initial states. For these two types of automata, first, we give the new definitions, then we analyze which results hold and which ones need to be adapted to the new concepts.

2.2. Cover Automata for Partial Words

A DFCA with "do not care" symbols, written $\diamond$-DFCA, is a cover automaton for the finite language $L_\diamond \subseteq \Sigma_\diamond^{\leq l}$. Please note that in a $\diamond$-DFA or $\diamond$-DFCA, it is not required to have for every state transitions with "do not-care" symbol $\diamond$. Thus, partial automata are usually presented as incomplete automata, namely, the transitions of "do not care" symbol to a dead state are omitted.

The language recognized by a $\diamond$-DFCA, A, over the extended alphabet $\Sigma \cup \{\diamond\}$ is $L' = \{w \mid \Sigma^\star \cup \{\diamond\} \mid \delta(q_0, w) \in F$ and $|w| \leq l\}$, where l is the length of the longest accepted word. We need to find the language over the original alphabet Σ, thus we apply a substitution $\sigma : \Sigma^\star \cup \{\diamond\} \longrightarrow 2^\Sigma$ to get $\sigma(L)$ as the σ-language over $\Sigma^\star$, accepted by the $\diamond$-DFCA.

In [13], as well as in [25], for the substitution σ we can have $\sigma(\diamond) = \Gamma \subseteq \Sigma$. In this paper, we only consider the case where $\sigma(\diamond) = \Sigma$, although most results are valid even if $\sigma(\diamond) \subset \Sigma$.

By replacing the "do not care" symbols in a $\diamond$-DFCA with all letters in Σ, the $\diamond$-DFCA becomes a NFCA. Thus, if L is a language accepted by a minimal $\diamond$-DFCA with n states, then the "do not care" state complexity of L is $dnccsc(L) = n$. Since any DFCA can be also considered a $\diamond$-DFCA, we have that $ncsc(L) \leq dnccsc(L) \leq dncsc(L) \leq sc(L)$.

3. Cover Automata with "Do Not Care" Symbols

In our study, we only need to see how "do not care" symbols influence state similarity and mergeability of two states, because everything that would be valid for NFCAs would then apply to $\diamond$-DFCAs. For strong mergeability, we always obtain deterministic transitions because we remove some of the states' transitions.

For a transition $t = p \xrightarrow{\alpha} q$, $p, q \in Q$, $\alpha \in \Sigma_\diamond$, and a substitution σ, we consider $\sigma(t) = \{p \xrightarrow{a} q | a \in \sigma(\alpha)\}$, i.e., the set of all transitions that can be obtained by substituting the letter $\alpha \in \Sigma_\diamond$ with $\sigma(\alpha)$. If $A = (Q, \Sigma_\diamond, \delta, q_0, F)$ is a DFCA for L, we can denote by $\Delta = \{p \xrightarrow{\alpha} q \mid \alpha \in \Sigma_\diamond, \delta(p, \alpha) = q\}$, i.e., the set of all transitions in the automaton A and $\Gamma = \{p \xrightarrow{a} q \mid (p \xrightarrow{a} q) \in \sigma(p \xrightarrow{\alpha} q), (p \xrightarrow{\alpha} q) \in \Delta)\}$, the set of all transitions obtained from the original ones by applying the substitution σ.

We now define the compatibility of two states in a $\diamond$-DFCA.

Definition 3. *Let $A = (Q, \Sigma_\diamond, \delta, q_0, F)$ be a $\diamond$-DFCA. Two states $p, q \in Q$ are σ-compatible for the substitution σ, denoted by $p \uparrow q$, if the set $\{(a, s) \mid (p \xrightarrow{a} s) \in \Gamma\} = \{(a, s) \mid (q \xrightarrow{a} s) \in \Gamma\}$.*

Two states $p, q \in Q$ are σ-strongly compatible, denoted $p \upuparrows q$, if they are σ-compatible, $\#\{s \mid (p \xrightarrow{a} s) \in \Gamma$ or $(q \xrightarrow{a} s) \in \Gamma\} \leq \#\Sigma_\diamond$, and if there are $s, r \in Q$ and $a \in \Sigma$ such that $(p \xrightarrow{a} s), (p \xrightarrow{a} r) \in \Gamma$, we either have $r = s$, or $(p \xrightarrow{b} s) \in \Delta$, for all $b \in \Sigma$, or $(p \xrightarrow{b} r) \in \Delta$, for all $b \in \Sigma$.

When the substitution σ is understood or in case $\sigma(\diamond) = \Sigma^\star$, it can be omitted and we say that p and q are compatible, respectively, strongly compatible.

In other words, two states are compatible if by applying the substitution of "do not care" symbols for the $\diamond$-transitions, we obtain the same destination states from p and q using the same letters in Σ. A weak merge, in the sense of Definition 2 can be used in case p and q are compatible, but we need to check that this procedure won't change the language.

At the same time, two states are strongly compatible if we can take a destination state s and all transitions with all letters to s can be replaced by only one transition using a "do not care" symbol, and all other destinations can be reached by at most one symbol from Σ for all the other transitions originating in p and q.

Thus, by merging state p with state q and considering the set T of consolidated transitions, we can replace transitions in T with transitions T' such that

1. we will have only one transition for each symbol in $\Sigma_\diamond$ and
2. by applying the substitution σ, we get the same consolidating transitions, i.e., $\sigma(T) = \sigma(T')$.

This new procedure can be defined for partial DFAs and it corresponds to the strongly merging procedure in Definition 2.

Let us check the time complexity required to:

1. Decide if two states are strongly compatible, and
2. Define a method to merge two strongly compatible states.

To decide if two states are strongly compatible, we need to check the following:

1. Check if $\{(a,s) \mid (p \xrightarrow{a} s) \in \sigma(p \xrightarrow{\alpha} s), p \xrightarrow{\alpha} s \in \Delta\} = \{(a,s) \mid (q \xrightarrow{a} s) \in \sigma(q \xrightarrow{\alpha} s), q \xrightarrow{\alpha} s \in \Delta\}$. If no, then the states are not strongly compatible, so we do not attempt to strongly merge them (Consolidate the outgoing transitions and modify them to get deterministic transitions only).
2. The number of destinations $\#\{s \mid (p \xrightarrow{a} s) \in \sigma(p \xrightarrow{\alpha} s), p \xrightarrow{\alpha} s \in \Delta\}$ can be at most $\#\Sigma_\diamond$. If not, the states are not strongly compatible.
3. If for a letter $a \in \Sigma$, there are at least three distinct states s_1, s_2, s_3, such that $(a, s_i) \in \{(a,s) \mid (p \xrightarrow{a} s) \in \sigma(p \xrightarrow{\alpha} s), p \xrightarrow{\alpha} s \in \Delta\}$, for all $1 \leq i < 3$, then the states are not strongly compatible.
4. If there exists a letter $a \in \Sigma$, such that if there are two states s_1, s_2 with $\{(a,s_1),(a,s_2)\} \subseteq \{(a,s) \mid (p \xrightarrow{a} s) \in \sigma(p \xrightarrow{\alpha} s), p \xrightarrow{\alpha} s \in \Delta\}$, then we must have either for all $b \in \Sigma \setminus \{a\}$, $b \in \sigma(\diamond)$, $(b, s_1) \in \{(a,s) \mid (p \xrightarrow{a} s) \in \sigma(p \xrightarrow{\alpha} s), p \xrightarrow{\alpha} s \in \Delta\}$, or for all $b \in \Sigma \setminus \{a\}$, $b \in \sigma(\diamond)$, $(b, s_2) \in \{(a,s) \mid (p \xrightarrow{a} s) \in \sigma(p \xrightarrow{\alpha} s), p \xrightarrow{\alpha} s \in \Delta\}$, but not both.
If this condition is not satisfied, we cannot replace all the transitions on $b \in \sigma(\diamond)$ to only one of the states s_1, or s_2 with the "do not care" symbol, and we do not obtain determinism for the transitions in the merged state.

Because $\#\{s \mid (p \xrightarrow{a} s) \in \sigma(p \xrightarrow{\alpha} s), p \xrightarrow{\alpha} s \in \Delta\} \leq \#\Sigma_\diamond$, all these steps, 1 to 4, take $O(1)$ time. Of course, for step 4, we may have two choices for the resulting automaton, but either one we choose, it takes constant time to do the merging. In step 4, if we have a transition from state p to state s with a letter $a \in \Sigma$ and a transition from state p to state s with "do not care" symbol $\diamond$, the transition from state p to state s with a letter $a \in \Sigma$ can be absorbed into transition from state p to state s with "do not care" symbol $\diamond$, as it is a redundant transition.

In Figure 1 are depicted all possible cases of merging two strongly compatible states, in case the alphabet is $\Sigma = \{a, b\}$.

Remark 1. *By strongly merging two states, we may obtain nondeterministic transitions. However, in the case of strongly compatible states, redundant transitions can be absorbed into the do not care symbol obtaining only deterministic ones.*

For defining the similarity relation in a $\diamond$-DFCA for two states p and q, we need the states to be similar in the corresponding NFCA, as in Definition 2 of [10].

Hence, we get the following:

Definition 4. *For a $\diamond$-DFCA $A = (Q, \Sigma_\diamond, \delta, q_0, F)$ two states p and q are similar, denoted $p \sim_A q$, if for all $w \in \Sigma^{\leq l - \max\{level_A(p), level_A(q)\}}$ and a partial word u such that $w \in \sigma(u)$, there is a partial word v, such that $w \in \sigma(v)$, and we also have that $\delta(p,u) \in F$ if $\delta(q,v) \in F$.*

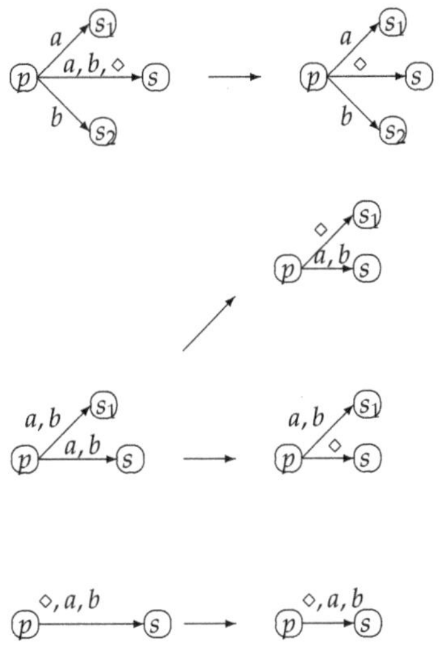

Figure 1. If the two states are strongly compatible and the result of strongly merging them is state p, the second $\diamond$ transition will be from p to s and it may overlap with an a or a b transition, or a diamond transition. In this case, we keep just one $\diamond$ transition and we drop the transitions on all letters $b \in \Sigma$ from p to s that are overlapping with it, to avoid nondeterminism. In all cases any non-deterministic choice can be avoided by "absorbing" a letter transition into the "do not-care" symbol transition.

Lemma 1. *Let $\diamond$-DFCA $A = (Q, \Sigma_\diamond, \delta, q_0, F)$ be a $\diamond$-DFCA. If two states p and q are similar, then they can be either strongly merged eliminating redundant transitions, if they are strongly compatible, or weakly merged otherwise.*

Proof. Let L' be the language of partial words accepted by A, $L = \sigma(L) \subseteq \Sigma^\star$ the associated finite language, and l the length of the longest words in L. Without any loss of generality, we may assume that $level_A(p) \leq level_A(q)$. Let $v \in \Sigma_\diamond^{\leq l}$ such that $\delta(q, v) \in F$, $v = \alpha v'$, $\alpha \in \Sigma_\diamond$. It follows that for every word $w \in \sigma(v)$ and any $x_q \in \sigma(y_q)$ such that $\delta(q_0, y_q) = q$ and $x_q w \leq l$, we have that $x_q w \in L$. We also have that $|x_q| \geq level(q) \geq level(p)$.

There is a partial word u, such that $w \in \sigma(u)$ and $\delta(p, u) \in F$ because $p \sim_A q$. Thus, by redirecting all transitions from q to p (the weakly merging method), we obtain a new automaton A' for which $x_q w$ is in the associated language of A'.

If we have a word w in the associated language of A', it means that there is a partial word z accepted by A' such that $w \in \sigma(z)$. If for every prefix of π of z, $\delta'(q_0, \pi) \neq p$, then $\delta(q_0, \pi) \neq p$, and z is accepted by A, therefore $w \in L$.

We have that $\delta'(p, v) \in F$ because $\delta'(q_0', z) \in F$ in case $\delta(q_0, \pi) = p$ for some π with $z = \pi v$. Since $p \sim_A q$, for $y \in \sigma(v)$, there is a partial word u, such that $y \in \sigma(u)$ and $\delta(q, u) \in F$. We have either $\delta(q_0, \pi) = p$, or $\delta(q_0, \pi) = q$ because A' is obtained from A by weakly merging q into state p. In both cases, $w \in \sigma(\pi u) \cup \sigma(\pi v)$, and either $\pi u \in L(A)$, or $\pi v \in L(A)$, so $w \in L$.

Hence, the language associated with the automaton A' does not change in case we do a weak merging of similar states.

If p and q are strongly compatible, let $w = aw'$, $a \in \sigma(\diamond)$. Thus, either $\delta'(p, a) = \delta(q, a)$, or $\delta'(p, \diamond) = \delta(q, a)$. Consequently, the word $x_q w$ is also in the language associated with the automaton A'.

If w is in the associated language of A', then there is a partial word z such that $w \in \sigma(z)$. In case $\delta(q_0, \pi) = p$ for some π with $z = \pi v$, because $\delta'(q_0', z) \in F$, then $\delta'(p, v) \in F$ and $w = xy$, $y \in \sigma(v)$.

Since $p \sim_A q$, for $y \in \sigma(v)$, there is a partial word u, such that $y \in \sigma(u)$, and $\delta(q,u) \in F$.

Because A' is obtained from A by strongly merging q into state p, we have either $\delta(q_0,\pi) = p$, or $\delta(q_0,\pi) = q$. In both cases, $w \in \sigma(\pi u) \cup \sigma(\pi v)$, and either $\pi u \in L(A)$, or $\pi v \in L(A)$, so $w \in L$. □

Let us see how we can use the above results to minimize ⋄-DFCAs.

In Section 2 of [12], the authors show that NFA minimization is NP-hard even in the case when the NFAs recognize finite languages, and they have limited non-determinism, i.e., the automata have at most one non-deterministic transition. Moreover, Corollary 7 on page 208 of [12], states that the minimization problem is NP-hard even if the input is given as a DFA. Their proof is based on the fact that the normal set cover problem is NP-complete [26,27]. Hence, if you consider these sets as paths, which corresponds to words, in an NFA, finding a minimal NFA is equivalent to finding a minimal set cover. For proving it in case of limited nondeterminism, they need a normal set cover B of a set C, i.e., for each $c \in C$, there is a subset B_c of B such that $c = \cup_{X \in B_c} X$ the elements in B_c are pairwise disjoint. The partition B is separable normal set basis for C if B can be written as a disjoint union of two other non-empty sets B_1 and B_2 such that for each $c \in C$, the subcollection B_c contains at most one element of B_1 and at most one element of B_2. To do that, they use a modified version of a known reduction from vertex cover to normal set basis (Lemma 4 in [28]), showing that the second problem is NP-hard. Using this result, they show that some instances of normal set basis sets in the partition will be pairwise disjoint and you can have just one state with two a-transitions. For (C,s), a separable normal set basis they consider, the language considered is $L = \{acb \mid c \in C, b \in c\}$, over a growing alphabet $\Sigma = \{a\} \cup \bigcup_{1 \le i \le n}\{c_i, b_{i_1}, \ldots, b_{i,n_i}\}$.

All accepted words are of length 3.

Therefore, for our case, we can use the same proof in two ways:

1. Either showing that ⋄-DFAs satisfy the conditions of Definition 1 page 201 of [12] and asking that the minimum length of the longest accepted string is at least 3, or
2. Use the same input as they use and replace the a symbol, that generates the nondeterministic transition, with a "do not care" symbol, so we get a ⋄-DFA. In this case, the only change would be that $L = \{\alpha cb \mid \alpha \in \Sigma, c \in C, b \in c\}$, and we would get several instances of the same problem, only with the first letter changed. Finding a minimal a normal set cover will only involve letters 2 and 3 for all paths from the start state to the final state, therefore, we can follow the same proof, but ignoring the part where they need to show that the minimal finite automaton is not ambiguous—in our case, that's not necessary. One can check that the proof works without any other change for ⋄-DFCAs, considering that the length of the longest accepted word is at least l, with $l \geq 3$.

It follows that:

Theorem 1. *Minimizing ⋄-DFCAs is NP-hard.*

Therefore, we need to seek simplification algorithms rather than minimization algorithms for ⋄-DFCAs. We already know [1,7,24,29] that all minimization algorithms for DFCAs are based on determining all similar states and merge them. For testing the similarity, one method [24,29], is to compute the gap function for two states p and q, where $gap(p,q)$ is the length of the shortest word that will distinguish between the states p and q. For ⋄-DFCAs, this means that we need to determine the length of the shortest word w such that:

1. if $w \in \sigma(u)$, $u \in \Sigma_\diamond^{\le l-\max\{level(p),level(q)\}}$, and $\delta(q,u) \in F$, then for any partial word $v \neq u, v \in \Sigma_\diamond^{\le l-\max\{level(p),level(q)\}}$ such that $w \in \sigma(v)$, we have that $\delta(p,v) \notin F$.

2. if $w \in \sigma(v)$, $v \in \Sigma_\diamond^{\leq l-\max\{level(p),level(q)\}}$, and $\delta(p,v) \in F$, then for any partial word $u \neq v$, $u \in \Sigma_\diamond^{\leq l-\max\{level(p),level(q)\}}$ such that $w \in \sigma(u)$, we have that $\delta(q,u) \notin F$.

It follows that:

1. if $\delta(p,a) = r$ and $\delta(q,a) = s$ and $gap(r,s) = k$, then $gap(p,q) \leq k+1$, and
2. if $\delta(p,a) = r$ and $\delta(q,\diamond) = s$, and $gap(r,s) = k$, then $gap(p,q) \leq k+1$.

Hence, we deduce that the gap function can be recursively computed as follows: $gap(p,q) = 1 + \min\{gap(r,s) \mid r = \delta(p,\alpha), s = \delta(q,\beta), \alpha,\beta \in \Sigma_\diamond$ and we have $\alpha = \beta$, or $\alpha \in \sigma(\beta)$, or $\beta \in \sigma(\alpha)\}$.

Because the number of transitions from p and q is bounded by $\#\Sigma_\diamond$, computing the gap function for a DFCA can be done in constant time for any pair p,q, if we know the gap function for all pairs r,s, such that $r = \delta(p,\alpha)$, $s = \delta(q,\beta)$, $\alpha,\beta \in \Sigma_\diamond$.

Two state would be then similar if $gap(p,q) \geq l - \max(level(p), level(q))$. With these observations and the fact that all known minimization algorithms for DFCAs are based on computing the similarity relation, we can modify any of the known minimizing algorithms for DFCAs [1,4,7,29] or l-DFCAs [24], to obtain a simplification algorithm for $\diamond$-DFCAs, without changing their computational complexity. Since the minimization algorithms for DFCAs are at most $O(n^4)$ [1,10], for all these simplification algorithms the time complexity will be at most $O(n^4)$ as well, which is better than the time complexity of the simplification algorithm proposed in [13]. Accordingly, we have obtained the following result:

Theorem 2. *For every DFCA minimization algorithm based on merging similar states having the run time complexity of $O(f(n))$, there is a simplification $\diamond$-DFCA algorithm having the same complexity of $O(f(n))$.*

A language L' is σ-minimal partial language for L if for any other language L'' such that $\sigma(L'') = L$, there is no word $w \in L'$ such we can find $x \in L''$ with x is weaker than w and $x \neq w$.

For example, $L' = \{\diamond a \diamond\}$ is σ-minimal partial language for $L = \{aaa, aab, baa, bab\}$. Indeed, if L'' is σ-partial language for L, $x \in L''$ and $w \in L'$ are such that $x \precsim w$, then we either have $x = \diamond\diamond\diamond$ or $x = w$. In the first case, $aba \in \sigma(L'') \neq L$ and in the second case $x \neq w$ is false. Therefore, L' is a σ-minimal partial language for L.

It must be noted that the simplification algorithm proposed in [13] obtains an approximation of the σ-minimal partial language L' for the regular language L, and obtaining this σ-minimal partial language is NP-hard [13]. The cover language L', for the finite language L that we obtain by applying the simplification algorithm, may not be a cover language for the σ-minimal partial language L_σ, i.e., $L' \cap \Sigma_\diamond^{\leq l} \neq L_\sigma$, but it is a σ-partial language that may have a lower state complexity than the original DFCA for L and $\sigma(L')$ is a cover language for L, i.e., $\sigma(L') \cap \Sigma^{\leq l} = L$. Please note that L_σ is the weakest partial language such that $\sigma(L_\sigma)$.

In Figure 2 for the language $L = \{a, b, aa, ab, ba, aaa, aab, aba, abb, baa, bab\}$, $\max\{|w| \mid w \in L\} = l = 3$, we have the partial language $L_1 = \{\diamond, ba, a\diamond, \diamond a\diamond, a\diamond\diamond\}$, recognized by the $\diamond$-automaton $A_1 = (\{a,b\}, \{0,1,2,3,4\}, \delta_1, 0, \{2,4\})$, with useful transitions $\delta_1(0,a) = 1$, $\delta_1(0,\diamond) = 2$, $\delta_1(0,b) = 3$, $\delta_1(1,\diamond) = 4$, $\delta_1(2,a) = 0$, $\delta(3,a) = 2$, $\delta_1(4,\diamond) = 2$, that is a σ-minimal partial cover language for L, i.e., $\sigma(L_1) \cap \Sigma^{\leq l} = L$, and the words in L_1 are the weakest possible with this property. We have that the $csc(L_1) = 5$.

The language L_2 recognized by the $\diamond$-automaton $A_2 = (\{a,b\}, \{0,1,2\}, \delta_2, 0, \{2\})$, with useful transitions $\delta_2(0,a) = 2$, $\delta_2(1,a) = 2$, $\delta_2(0,b) = 1$, $\delta_2(2,\diamond) = 2$, is a σ-partial cover language for L, i.e., $\sigma(L_2) \cap \Sigma^{\leq l} = L$, and $csc(L_2) = 3$. $L_2 \cap \Sigma_\diamond^{\leq 3} = \{a, b, ba, a\diamond, a\diamond\diamond, ba\diamond\}$ and it contains 6 words, but $L_1 \cap \Sigma^{\leq 3} = \{\diamond, ba, a\diamond, a\diamond\diamond, \diamond a\diamond\}$ has only 5 words.

The example above shows that it is impossible to obtain a cover language for the σ-partial minimal language that has, at the same time, the minimal cover state complexity.

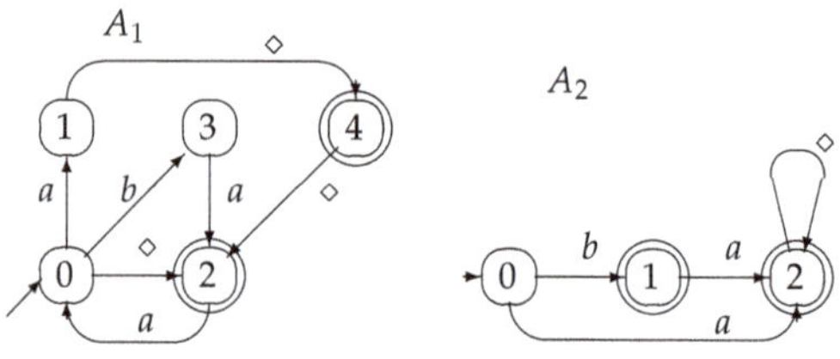

Figure 2. Automaton A_1, left, and automaton A_2, right. $L(A_1)$ is a cover language for the σ-minimal partial language for L, and $L(A_2)$ is a cover partial automaton for L, where $L = \{a, b, aa, ab, ba, aaa, aab, aba, abb, baa, bab\}$.

4. Multiple Entry DFCAs

For multiple entry DFCAs, we can have two possible flavors of extensions. The first one and the easiest to consider is the same maximum length for all words accepted by the m-DFCA. The second approach is to consider for each initial state a different maximum length. Therefore, we can use the following definition:

Definition 5. *A multiple entry DFCA with m initial states, i.e., an m-DFCA, is a structure $A = (Q, \Sigma, \delta, Q_0, F, \Lambda)$, such that Q, Σ, δ, and F are the same as for usual DFCAs, Q_0 is the set of initial states, $\#Q_0 = m$, and $\Lambda = (l_1, \ldots, l_m)$ is a sequence of m integers representing the maximum accepted length for each initial state. If $Q_0 = \{q_{0,1}, \ldots q_{0,m}\}$ and $\Lambda = (l_1, l_2, \ldots, l_m)$, the language accepted by the m-DFCA A is*

$$L(A) = \{x \in \Sigma^\star \mid \delta(q_{0,i}, x) \in F \text{ and } |x| \leq l_i, \text{ for some } 1 \leq i \leq m\}.$$

We have the condition $L(A) = \cup_{i=1}^m L_i$, where $L_i = L(A_i) \cap \Sigma^{\leq l_i}$, $A_i = (Q, \Sigma, \delta, q_{0,i}, F)$. The automata A_i are subautomata induced by the m-DFCA A.

We observe from the above definition that the set of initial states has the size m. Thus, by replacing the set Q_0 with an m-tuple $Q_0 = (q_{0,1}, \ldots q_{0,m})$, if we assign two possible lengths to an initial state, it does not change the accepted language. Assume that the initial states $q_{0,i}$ and $q_{0,j}$ are the same, i.e., $q_{0,i} = q_{0,j}$. The automaton will then accept all the words of length less than $\max\{l_i, l_j\}$, meaning that we can eliminate the one with the lowest maximum length, getting an $(m-1)$-DFCA.

In the example given by Björklund and Martens [12], they prove that minimizing m-DFAs is NP-hard, using a construction with a finite language. Because any m-DFA for a finite language is also an m-DFCA, by setting l_i to be the length of the longest accepting walk starting at $q_{0,i}$, it follows that

Theorem 3. *Minimizing m-DFCA is an NP-hard problem.*

We can reduce the size of m-DFCA efficiently in a similar way to the previous case for partial automata, obtaining a simplified m-DFCA by merging states. To avoid changing the language recognized by an m-DFCA A, the simplest solution is to merge similar states in all the subautomata A_i with the corresponding maximum length l_i. Any other merge of states will modify at least one of the languages involved, which will not guarantee that their union stays the same as before. Therefore, we can obtain the following definition for similarity in m-DFCAs:

Definition 6. *Let $A = (Q, \Sigma, \delta, Q_0, F, \Lambda)$ be an m-DFCA for the finite language L. Two states p and q are similar if p and q are similar in all cover automata $A_i = (Q, \Sigma, \delta, q_{0,i}, F, \lambda_i)$, $1 \leq i \leq m$.*

The simplification algorithms for m-DFCAs can be obtained as before by modifying existing DFA-minimization algorithms, therefore in what follows we will focus on state complexity problem, namely, on constructing a minimal DFCA for the same language, and evaluating the state complexity of this transformation.

It is known that for NFA to DFA transformation for finite languages, [30], in case of a binary alphabet, the upperbound is $2^{\frac{n}{2}+1} - 1$ if n is even, and $3 \cdot 2^{\frac{n-1}{2}} - 1$, if n is odd, n being the number of states of the NFA. Moreover, it is reached by the language $L_{m,n} = \{a,b\}^{\leq m}a\{a,b\}^n$ when $n = m$, or $n = m-1$, so $sc(L_{m-1,m-1}) = 2^{m+1} - 1$, and $sc(L_{m,m-1}) = 3 \cdot 2^m - 1$.

Because the length of the longest word is $m+n+1$, the minimal nondeterministic finite automaton recognizing this language will have at least $m+n+2$ states, while the minimal DFA will have at least $m+n+3$ states, [30], Theorem 2. A minimal $(m+1)$-DFA for $L_{m,n}$ has the same number of states as the NFA, plus the sink state, therefore, is minimal as a m-DFA. A minimal $(m+1)$-DFCA for $L_{m,n}$ has the same number of states as the NFA minus one, because the sink state is similar with state 0, and that is the only possible similarity. We can see that the minimal nondeterministic cover automaton for $L_{m,n}$ has only $n+2$ states. For this NFCA for $L_{m,n}$, the initial state is obtained by merging all the $m+1$-entries into one, so a minimal $(m+1)$-DFCA must have m states more than the NFCA, that is $m+n+1$.

Starting from an m-DFA for a finite language, we can construct an equivalent NFA by observing that there is one initial state $q_{0,0}$ with no incoming transition, and for each initial state $q_{0,i} \in Q_0$, if $\delta(q_{0,i}, a) = s_i$, we can add the transitions from $q_{0,0}$ with a in s_i, and we delete the sink state. This way, for the m-DFA to DFA transformation, we obtain the limits for NFA to DFA transformation, but we need to consider the extra sink state for m-DFAs.

Therefore, we just proved the following result:

Theorem 4. *In case of a binary alphabet, the upperbound for a n-state m-DFCA to DFCA transformation is* $2^{\frac{n-1}{2}+1} - 1$ *if n is odd, and* $3 \cdot 2^{\frac{n-2}{2}} - 1$, *if n is even, and the bound is reached.*

Figure 3 shows a 5-DFA with 11 states for the language $L_{4,4}$. This 5-DFA is a minimal multi-entry DFA. Any nondeterministic finite automaton recognizing $L_{4,4}$ must have at least 11 states which is the length of the longest word in $L_{4,4}$ plus 1. A DFA, multi-entry or not must have a sink state, because the language is finite, therefore the automaton depicted in Figure 3a is minimal. The corresponding multi-entry DFCA, Figure 3b has the dead state d similar with state 0, so we can reduce the size by one state. The NFCA in Figure 3c recognizes $L_{4,4}$, and it has only 6 states. The general case for $L_{m,n-1}$ is depicted in Figure 4.

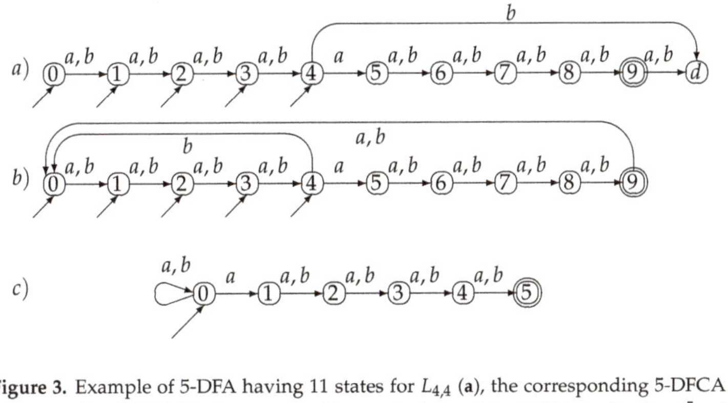

Figure 3. Example of 5-DFA having 11 states for $L_{4,4}$ (**a**), the corresponding 5-DFCA (**b**), and the corresponding NFCA (**c**). A corresponding equivalent minimal DFA for $L_{4,4}$ has $2^5 - 1$ states and a minimal DFCA has 16 states.

The upperbound for m-DFA to DFCA transformation is the same as the NFA to DFCA transformation, but there is one difference. We have to consider that in the m-DFA, we must have one more state as dead state because the language is finite, while in the NFA that state can be eliminated, as it is not useful.

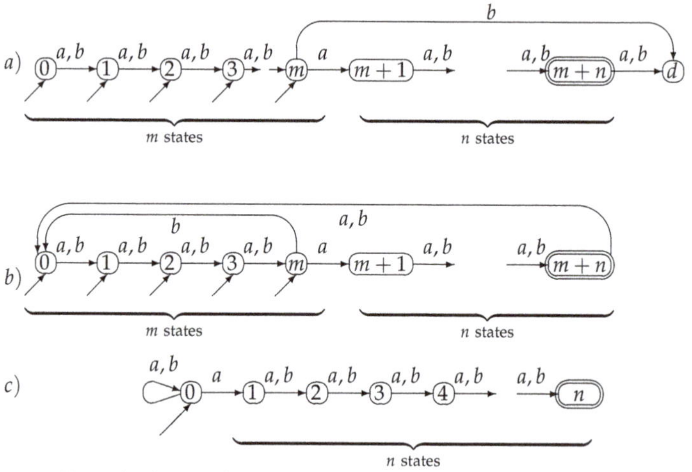

Figure 4. Example of m-DFA having $m + n + 2$ states for $L_{m,n-1}$ (**a**), the corresponding m-DFCA (**b**), and the corresponding NFCA (**c**).

5. Discussion

This paper investigates the feasibility of extending the definition of cover automata to include the cases when we allow multiple entries or "do not care" symbols. Because both operations induce a degree of nondeterminism, existing minimization algorithms working on DFAs may not give the smallest automaton in the new class, so we need to verify their run time complexity, and their correctness.

The previous automata constructions prove that minimization problems for certain nondeterministic automata are NP-hard. We checked that the same examples could also be used without any significant modification for cover automata. Hence, the results hold. The proof details for these results were omitted, as it can be found in [12] for NFAs and m-DFAs. For cover automata, we only needed to add the "do not care" symbols to substitute one letter in one transition for the first case and to add the maximum length for both cases.

In the previous studies [31,32] on state complexity of partial words DFAs we can find particular classes of languages where minimization bounds can be established. The general case is still open.

I proved that there are simplification algorithms with the same time complexity as existing minimizing DFCA algorithms. I have also computed the state complexity bound for m-DFCA to DFCA transformation.

In the case of DFCAs, the idea was floating around even in the 1960s [33,34], but no formal definition was given until 1998. That is the reason why until 2001, there was no result published on this topic, but several papers followed after the publication of [1]. In this paper, I give the required formal definitions for two DFCA extensions, and I also prove some essential results necessary to start any further investigation.

There are several questions that one may ask; for example, the following questions might be of interest:

1. Finding the state complexity of operations on $\diamond$-DFCAs.
2. Finding the state complexity of operations on with multiple entry DFCA.
3. Considering or exclusive nondeterministic finite cover automata, XNFCA.
4. Considering multiple entry XDFCA.
5. State complexity of XOR-star, XOR-concatenation for finite languages.
6. State complexity of XOR-reverse. Algebraic properties of finite languages and XOR acceptance—same length and different lengths.

7. Considering multiple lengths for multiple entry DFAs.

The study of state complexity of operations with finite languages represented by finite deterministic automata was started in [35] and later on in [36]. Later on, the state complexity of operations using nondeterministic automata were considered in [5,37–39]. It would be interesting to see where would the state complexity bounds for operations on these extensions that introduce a low level of nondeterminism would fit: closer to the deterministic results or closer to the non-deterministic models.

For each of these extensions, we need to study the following aspects:

1. Analyze the complexity of the membership problem.
2. Investigate the existence of complexity of minimization/simplification algorithms.
3. Find and evaluate the dynamic complexity and state complexity of transforming the new automata model into a known one.
4. Find bound for state complexity of operations done using the new representation model.

6. Conclusions

Two extensions are formally defined for the cover automata model of representing finite languages. Fundamental properties of these extensions were checked and proved, followed by the methodology at the end of Section 5. This article also proves that the minimization of $\diamond$-DFCAs and m-DFCAs is NP-hard, and it shows a process for obtaining simplification algorithms based on merging states in Theorem 2. The upper bound for m-DFCA to DFCA transformation is computed and proved in Theorem 4. The Discussion section also includes open problems and future research directions.

Funding: This research received no external funding.

Acknowledgments: I am very grateful to the organizers for this volume, In Memoriam of emeritus Professor Solomon Marcus. I was fortunate to participate in the celebration of his 90th birthday when we reconnected after many years, as he was one of the most influential professors in my career. I had the honor of having him as one of the professors during my time as a student at the University of Bucharest. Professor Solomon Marcus will be greatly missed for his friendly and enthusiastic personality by all of us.

Conflicts of Interest: The author declares no conflict of interest.

References

1. Câmpeanu, C.; Sântean, N.; Yu, S. Minimal Cover Automata for Finite Languages. In Proceedings of the Third International Workshop on Implementing Automata (WIA 1998), Rouen, France, 17–19 September 1998; pp. 33–42.
2. Câmpeanu, C.; Santean, N.; Yu, S. Minimal Cover-Automata for Finite Languages. *Theor. Comput. Sci.* **2001**, *267*, 3–16. [CrossRef]
3. Câmpeanu, C.; Păun, A. Counting the Number of Minimal DFCA Obtained by Merging States. *IJFCS* **2003**, *14*, 995–1006. [CrossRef]
4. Câmpeanu, C.; Paun, A.; Smith, J.R. Incremental construction of minimal deterministic finite cover automata. *Theor. Comput. Sci.* **2006**, *363*, 135–148. [CrossRef]
5. Gao, Y.; Moreira, N.; Reis, R.; Yu, S. A Survey on Operational State Complexity. *arXiv* **2015**, arXiv:1509.03254.
6. Gruber, H.; Holzer, M.; Jakobi, S. More on deterministic and nondeterministic finite cover automata. *Theor. Comput. Sci.* **2017**, *679*, 18–30. [CrossRef]
7. Körner, H. A Time and Space Efficient Algorithm for Minimizing Cover Automata for Finite Languages. *Int. J. Found. Comput. Sci.* **2003**, *14*, 1071–1086. [CrossRef]
8. Wolfsteiner, S. Grammatical Complexity of Finite Languages. Ph.D. Thesis, TU Wien, Vienna, Austria, 2020.
9. Câmpeanu, C. Simplifying Nondeterministic Finite Cover Automata. In Proceedings of the 14th International Conference on Automata and Formal Languages, AFL 2014, Szeged, Hungary, 27–29 May 2014; Volume 151, pp. 162–173. [CrossRef]
10. Câmpeanu, C. Nondeterministic Finite Automata. In *Scientific Annals of Cuza University*; Alexandru Ioan Cuza University: Iaşi, Romania, 2015; pp. 1–25.
11. Fischer, M.J.; Paterson, M.S. String-matching and other products. *Complex. Comput.* **1974**, *7*, 113–126.
12. Björklund, H.; Martens, W. The tractability frontier for NFA minimization. *J. Comput. Syst. Sci.* **2012**, *78*, 198–210. [CrossRef]
13. Blanchet-Sadri, F.; Goldner, K.; Shackleton, A. Minimal partial languages and automata. In Proceedings of the 19th International Conference on Implementation and Application of Automata, Giessen, Germany, 30 July–2 August 2014; Springer: Berlin/Heidelberg, Germany, 2014; pp. 110–123.

14. Marcus, S. Personal Communication with the Ocasion of His 90th Birthday, 2015.
15. Muthukrishnan, S.; Palem, K. Non-Standard Stringology: Algorithms and Complexity. In Proceedings of the Twenty-Sixth Annual ACM Symposium on Theory of Computing, Montreal, QC, Canada, 23–25 May 1994; Association for Computing Machinery: New York, NY, USA, 1994; pp. 770–779. [CrossRef]
16. Holzer, M.; Jakobi, S.; Wendlandt, M. On the computational complexity of partial word automata problems. In *IFIG Research Report 1404*; Institut für Informatik, Justus-Liebig-Universitäat Gießen: Gießen, Germany, 2014; pp. 1–27.
17. Holzer, M.; Jakobi, S.; Wendlandt, M. On the computational complexity of partial word automata problems. In Proceedings of the Sixth Workshop on Non-Classical Models for Automata and Applications—NCMA 2014, Kassel, Germany, 28–29 July 2014; Bensch, S., Freund, R., Otto, F., Eds.; Österreichische Computer Gesellschaft: Wien, Austria, 2014; Volume 304, pp. 131–146.
18. Galil, Z.; Simon, J. A note on multiple-entry finite automata. *J. Comput. Syst. Sci.* **1976**, *12*, 350–351. [CrossRef]
19. Holzer, M.; Salomaa, K.; Yu, S. On the State Complexity of k-Entry Deterministic Finite Automata. *J. Autom. Lang. Comb.* **2001**, *6*, 453–466.
20. Polák, L. Remarks on Multiple Entry Deterministic Finite Automata. *J. Autom. Lang. Comb.* **2007**, *12*, 279–288.
21. Veloso, P.A.; Gill, A. Some remarks on multiple-entry finite automata. *J. Comput. Syst. Sci.* **1979**, *18*, 304–306. [CrossRef]
22. Hopcroft, J.E.; Motwani, R.; Ullman, J.D. *Introduction to Automata Theory, Languages and Computation*; PearsonEd/AW: Boston, UK; San Francisco, CA, USA; New York, NY, USA; Toronto, Canada; Montreal, Canada, 2007.
23. Gill, A.; Kou, L.T. Multiple-entry finite automata. *J. Comput. Syst. Sci.* **1974**, *9*, 1–19. [CrossRef]
24. Jez, A.; Maletti, A. Computing All L-Cover Automata Fast. In Proceedings of the 16th International Conference on Implementation and Application of Automata, Blois, France, 13–16 July 2011; Springer: Berlin/Heidelberg, Germany, 2011; pp. 203–214.
25. Dassow, J.; Manea, F.; Mercas, R. Connecting Partial Words and Regular Languages. In *How the World Computes—Turing Centenary Conference and 8th Conference on Computability in Europe, CiE 2012, Cambridge, UK, 18–23 June 2012*; Cooper, S.B., Dawar, A., Löwe, B., Eds.; Springer: Berlin/Heidelberg, Germany, 2012; Volume 7318, pp. 151–161. [CrossRef]
26. Karp, R.M., Reducibility among Combinatorial Problems. In *Complexity of Computer Computations: Proceedings of a symposium on the Complexity of Computer Computations, Held 20–22 March 1972, at the IBM Thomas J. Watson Research Center, Yorktown Heights, New York, and Sponsored by the Office of Naval Research, Mathematics Program, IBM World Trade Corporation, and the IBM Research Mathematical Sciences Department*; Springer: Boston, MA, USA, 1972; pp. 85–103. [CrossRef]
27. Cormen, T.H.; Leiserson, C.E.; Rivest, R.L.; Stein, C. *Introduction to Algorithms*, 3rd ed.; MIT Press: Cambridge, MA, USA, 2009.
28. Jiang, T.; Ravikumar, B. NFA Minimization problems are Hard. *SIAM J. Comput.* **1993**, *22*, 117–141. [CrossRef]
29. Câmpeanu, C.; Paun, A.; Yu, S. An Efficient Algorithm for Constructing Minimal Cover Automata for Finite Languages. *Int. J. Found. Comput. Sci.* **2002**, *13*, 83–97. [CrossRef]
30. Salomaa, K.; Yu, S. NFA to DFA transformation for finite languages over arbitrary alphabets. *J. Aut. Lang. Comb.* **1997**, *2*, 177–186.
31. Blanchet-Sadri, F.; Goldner, K.; Shackleton, A. Minimal partial languages and automata. *RAIRO Theor. Inform. Appl.* **2017**, *51*, 99–119. [CrossRef]
32. Balkanski, E.; Blanchet-Sadri, F.; Kilgore, M.; Wyatt, B. On the state complexity of partial word DFAs. *Theor. Comput. Sci.* **2015**, *578*, 2–12. Implementation and Application of Automata, Revised Selected Papers. [CrossRef]
33. Gold, E.M. Language identification in the limit. *Inf. Control* **1967**, *10*, 447–474. [CrossRef]
34. Yu, S. Cover Automata for Finite Languages. *EATCS Bull.* **2007**, *92*, 65–74.
35. Maslov, A.N. Estimates of the number of states of finite automata. *Dokl. Akad. Nauk SSSR* **1970**, *194*, 1266–1268 (Russian). English Translation: *Sov. Math. Dokl.* **1970**, *11*, 1373–1375.
36. Yu, S.; Zhuang, Q.; Salomaa, K. The State Complexities of Some Basic Operations on Regular Languages. *Theor. Comput. Sci.* **1994**, *125*, 315–328. [CrossRef]
37. Holzer, M.; Kutrib, M. State Complexity of Basic Operations on Nondeterministic Finite Automata. *Lect. Notes Comput. Sci.* **2003**, *2608*, 148–157.
38. Holzer, M.; Kutrib, M. Unary Language Operations and Their Non-deterministic State Complexity. *Lect. Notes Comput. Sci.* **2003**, *2450*, 162–172.
39. Holzer, M.; Kutrib, M. Nondeterministic Finite Automata—Recent Results on the Descriptional and Computational Complexity. *Int. J. Found. Comput. Sci.* **2009**, *20*, 563–580. [CrossRef]

axioms

MDPI

Article

A Hypergraph Model for Communication Patterns

Gabriel Ciobanu

Faculty of Computer Science, Alexandru Ioan Cuza University, 700506 Iasi, Romania; gabriel@info.uaic.ro
† In Memoriam Solomon Marcus (1925–2016).

Abstract: The article deals with interaction in concurrent systems. A calculus able to express specific communication patterns is defined, together with its abstract control structures. A hypergraph model for these structures is presented. The hypergraphs are able to properly express the communication patterns, providing a fully abstract model for the pattern calculus. It is also proved that the hypergraph model preserves the operational reductions of processes from pattern calculus and of the actions from the control structures.

Keywords: process calculus; communication patterns; control structures; hypergraph model

Citation: Ciobanu, G. A Hypergraph Model for Communication Patterns. *Axioms* **2022**, *11*, 8. https://doi.org/10.3390/axioms11010008

Academic Editor: Cristian S. Calude

Received: 30 October 2021
Accepted: 2 December 2021
Published: 23 December 2021

1. Introduction

Mathematics has sometimes been called a 'science of patterns' [1], meaning that patterns are at the heart of mathematics. The nice description of mathematics as the "study of patterns" was given by G.H. Hardy in his book *A Mathematician's Apology*: "A mathematician, like a painter or a poet, is a maker of patterns. If his patterns are more permanent than theirs, it is because they are made with ideas". Essentially, patterns are regularities that we can perceive. Regarding the skilful ability of applying a pattern to multiple contexts, Solomon Marcus was a wonderful example of applying knowledge patterns to surprising contexts.

Since mathematics and technology have developed a fruitful relationship over past few decades, patterns have been investigated recently in modern fields. New communication technologies have changed the computing landscape, and the Internet is now a platform for large scale distributed programming. Nowadays, we deal with global computation based on multiple interactions with the environment (instead of isolated systems). Concurrency is essential now in computer science; web servers handle multiple simultaneous clients, and cloud servers allow several simultaneous applications and users. Message-passing represents a way in which concurrent processes communicate (a process is an instance of a running program). In software architecture, a messaging pattern describes how two different processes communicate with each other. In telecommunications, a message exchange pattern describes the messages required by a communications protocol or the message flow between parties involved in communication. For example, when navigating on Internet (representing the channel), a web browser (the communicating party) uses the communication protocol HTTP to request a web page from the server (another communicating party). In general, the interaction between clients and servers follows a specific communication pattern: the client sends a request, the server returns a response, and so on. Such an exchange of messages is only an example of communication patterns. More complicated behaviours appear due to the concurrent interaction of communicating processes; this complexity reveal the necessity to find new ways to describe and build up concurrent systems. The communicating parties in such a concurrent system should have a common language to communicate; moreover, they should follow the rules defined in a communications protocol. There exits already a calculus able to express communication patterns called join calculus; it is used as a basis for some programming languages (JoCaml and Cω), but also as a basis for libraries (embedded in C#, F# and Scala). This calculus is based on 'join patterns', namely rules describing how a certain combination of values sent through multiple channels triggers a specific reaction and removes the values from the communication channels. Interaction in such a calculus is provided by sharing the communication channels names.

We introduce the pattern calculus as a weak version of join calculus. After presenting the control structures for pattern calculus, we provide a hypergraph model of these structures. The hypergraphs are able to express properly the communication patterns described by the pattern calculus. It is proved that the hypergraph model is fully abstract for the calculus; a model is fully abstract if all observationally equivalent processes represent the same object in the model, meaning also that processes with different behaviour are not mapped to the same hypergraph. Furthermore, there is a correspondence between the dynamics of the processes of the calculus and their hypergraph representation.

2. Pattern Calculus

The pattern calculus is inspired by the join calculus [2], a calculus proposed to underlie programming languages for distributed systems. A presentation of the join calculus can also be found in [3]. The specific construction of the new calculus is the definition of communication channels: $\mathtt{def}\ u\langle y\rangle \triangleright P\ \mathtt{in}\ Q$. To elucidate the simplicity of this syntactical construction, let us say that it could be expressed in the π-calculus [4] by using several syntactical constructions: $\mathtt{def}\ u\langle y\rangle \triangleright P\ \mathtt{in}\ Q = \nu u.(!u(y).P \mid Q)$.

The syntax of the pattern calculus is defined by using a countable set $\mathcal{X}$ of *names* ranging over $u, v, x, \ldots$, together with $\tilde{u}, \tilde{v}, \tilde{x}, \tilde{y} \ldots$ ranging over finite strings of names. We use $P, Q, R, \ldots$ ranging over the set of processes. The set $\mathcal{P}$ of processes contains an *empty process* 0, as well as an *output message* $u\langle v\rangle$ sending v by using a channel u. The process $P \mid Q$ describes the *parallel composition* of processes P and Q. The communication between processes is achieved by the *channel definition* $\mathtt{def}\ u\langle v\rangle \triangleright P\ \mathtt{in}\ Q$ indicating that the interaction of processes P and Q is realized by the channel u (which is created only for the communication between them).

Definition 1. *The processes of our calculus are defined by the following syntax:*

$$P ::= 0 \mid u\langle v\rangle \mid P \mid Q \mid \mathtt{def}\ u\langle v\rangle \triangleright P\ \mathtt{in}\ Q\,.$$

In $\mathtt{def}\ u\langle v\rangle \triangleright P\ \mathtt{in}\ Q$, both u and v are bound. The scope of v is P, while the scope of u is given by the whole definition. It is worth noting that only this definition binds names. The *free names* are defined inductively by $\mathtt{fn}(0) = \emptyset$, $\mathtt{fn}(u\langle v\rangle) = \{u, v\}$, $\mathtt{fn}(P \mid Q) = \mathtt{fn}(P) \cup \mathtt{fn}(Q)$, and $\mathtt{fn}(\mathtt{def}\ u\langle v\rangle \triangleright P\ \mathtt{in}\ Q) = (\mathtt{fn}(Q) \cup (\mathtt{fn}(P) - \{v\})) - \{u\}$.

A substitution $\{y/x\}P$ replaces all the free occurrences of name x in P by name y; name-capture is avoided by using the α-conversion (defined in the standard way).

Definition 2. *A* structural congruence $\equiv\ \subseteq \mathcal{P} \times \mathcal{P}$ *is defined as the smallest congruence satisfying the following axioms:*

- $\mathtt{def}\ u\langle v\rangle \triangleright P\ \mathtt{in}\ Q \equiv \mathtt{def}\ u\langle t\rangle \triangleright \{t/v\}P\ \mathtt{in}\ Q$, *if* $t \notin \mathtt{fn}(P)$
- $\mathtt{def}\ u\langle v\rangle \triangleright P\ \mathtt{in}\ Q \equiv \mathtt{def}\ w\langle v\rangle \triangleright \{w/u\}P\ \mathtt{in}\ \{w/u\}Q$ *if* $v \notin \{u, w\}$ *and* $w \notin \mathtt{fn}(P \mid Q)$
- $Q_1 \mid \mathtt{def}\ u\langle v\rangle \triangleright P\ \mathtt{in}\ Q_2 \equiv \mathtt{def}\ u\langle v\rangle \triangleright P\ \mathtt{in}\ (Q_1 \mid Q_2)$ *if* $u \notin \mathtt{fn}(Q_1)$
- $\mathtt{def}\ u\langle v\rangle \triangleright P_1\ \mathtt{in}\ \mathtt{def}\ w\langle t\rangle \triangleright P_2\ \mathtt{in}\ Q \equiv \mathtt{def}\ w\langle t\rangle \triangleright P_2\ \mathtt{in}\ \mathtt{def}\ u\langle v\rangle \triangleright P_1\ \mathtt{in}\ Q$ *if* $u \neq w$, $u \notin \mathtt{fn}(P_2)$, *and* $w \notin \mathtt{fn}(P_1)$
- $P \mid 0 \equiv P \qquad P \mid Q \equiv Q \mid P \qquad (P \mid Q) \mid R \equiv P \mid (Q \mid R)$.

The evolution of the processes is given by a reduction relation $\rightarrow$.
For the specific construct $\mathtt{def}\ u\langle y\rangle \triangleright P\ \mathtt{in}\ Q$, the reduction is described mainly by:

$$\mathtt{def}\ u\langle y\rangle \triangleright P\ \mathtt{in}\ Q \mid u\langle v\rangle \ \rightarrow\ \mathtt{def}\ u\langle y\rangle \triangleright P\ \mathtt{in}\ Q \mid \{v/y\}P.$$

More exactly, process Q can send a name v along the channel u, while process P waits at the other end of channel u to receive certain channel names. When the name v is received, process Q continues its execution in parallel with process P in which all free occurrences of y are replaced by v, i.e., $\{v/y\}P$. Then, channel u remains open to receive other names. The formal definition is given below.

Definition 3. *The* reduction relation $\rightarrow \subseteq \mathcal{P} \times \mathcal{P}$ *is defined as the smallest relation satisfying:*

r1: $\texttt{def}\ u_1\langle v_1\rangle \triangleright Q_1\ \texttt{in}\ \texttt{def}\ u_2\langle v_2\rangle \triangleright Q_2\ \texttt{in}\ \ldots\ \texttt{def}\ u_n\langle v_n\rangle \triangleright Q_n\ \texttt{in}\ (P \mid u_i\langle v\rangle) \rightarrow$
$\texttt{def}\ u_1\langle v_1\rangle \triangleright Q_1\ \texttt{in}\ \texttt{def}\ u_2\langle v_2\rangle \triangleright Q_2\ \texttt{in}\ \ldots\ \texttt{def}\ u_n\langle v_n\rangle \triangleright Q_n\ \texttt{in}\ (P \mid \{v/v_i\}Q_i)$
if $\{u_{i+1},\ldots,u_n\} \cap (\texttt{fn}(Q_i) \cup \{u_i\}) = \emptyset$ *with* $i \in [n]$ *and* $n \geq 1$

r2: $$\frac{P_1 \rightarrow P_2}{\texttt{def}\ u\langle v\rangle \triangleright Q\ \texttt{in}\ P_1 \rightarrow \texttt{def}\ u\langle v\rangle \triangleright Q\ \texttt{in}\ P_2}$$

r3: $$\frac{P_1 \equiv Q_1,\ Q_1 \rightarrow Q_2 \text{ and } Q_2 \equiv P_2}{P_1 \rightarrow P_2}.$$

It is worth noting that there is no rule for parallel composition in the definition of this reduction. The following (easy-to-prove) results show that such a rule for parallel composition is just a consequence.

Lemma 1. *Considering any substitution* $\sigma = \{x/y\}$,
$$P \equiv Q \text{ implies } \sigma P \equiv \sigma Q, \text{ and } P_1 \rightarrow P_2 \text{ implies } \sigma P_1 \rightarrow \sigma P_2.$$

Proposition 1. $P_1 \rightarrow P_2$ *implies* $Q \mid P_1 \rightarrow Q \mid P_2$.

3. Hypergraphs

In the theory of distributed systems, Petri nets [5] and π-nets [6] provide both algebraic and graphical descriptions for concurrent systems. Compared to Petri nets, our hypergraph model for pattern calculus has a flexible structure; compared to π-nets, the hypergraph model is simpler, but still able to describe a large class of processes.

Following [7], we present the definitions for hypergraphs and some standard related notions such as isomorphism and contractions (on nodes and on edges). For a set S of *hyperedges* and a set V of *vertices*, it is defined an *incidence* relation $E \subseteq S \times V$. A *rooted hypergraph* is a tuple $H = \langle S, V, E, s\rangle$, where $s \in S$ is the *root hyperedge*. For a hypergraph H, we use the notations S_H, V_H, E_H and s_H.

The graphical presentation of a hypergraph H is provided by:

- A hyperedge t represented as an oval with its name t outside.
- A vertex v represented as a point having the name v.
- An element (t, v) of the incidence relation represented as lines from the hyperedge t to the vertex v; "v lies on t" whenever $(t, v) \in E_H$.
- The root indicated by an arrow pointing to the hyperedge s_H.

Example 1. *Considering a hypergraph with* $S_H = \{s, t, t'\}$, $V_H = \{v, w, w'\}$, $E_H = \{(s, v), (t, v), (t, w), (t', v), (t', w')\}$ *and* $s_H = \{t\}$, *various graphical representations are depicted in Figure 1. The left representation uses only lines of non-zero length, while the right one uses only lines of zero length. The representation in the middle is a compromise (between lengths) to provide a reasonable picture.*

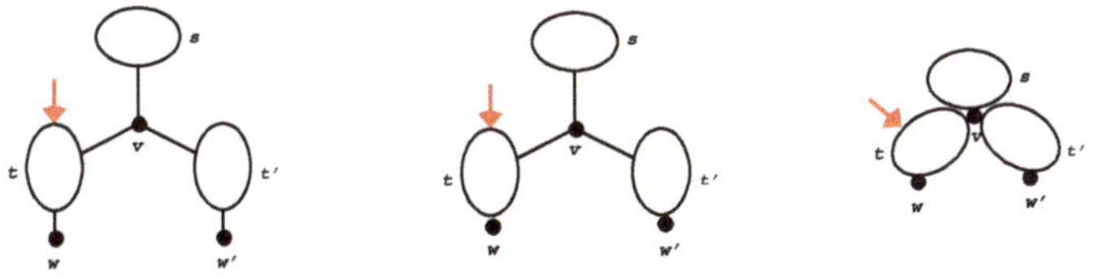

Figure 1. Graphical representations of a hypergraph.

For a given rooted hypergraph H and a nonempty subset $W \subseteq V_H$ of nodes, a *contraction on vertices* is specified by the hypergraph H/W with the same root hyperedge

($s_{H/W} = s_H$), the same hyperedges ($S_{H/W} = S_H$), but with $V_{H/W} = (V_H \setminus W) \cup \{v\}$ for a fresh $v \notin V_H$ and $E_{H/W} = (E_H \setminus S_H \times W) \cup \{ (t,v) \mid \{t\} \times W \cap E_H \neq \emptyset \}$.

For a given rooted hypergraph H and a nonempty subset $T \subseteq S_H$ of hyperedges, a *contraction on hyperedges* is specified by the hypergraph H/T with the same set of vertices ($V_{H/T} = V_H$), but with $S_{H/T} = (S_H \setminus T) \cup \{t\}$ for a fresh $t \notin S_H$, and $E_{H/T} = (E_H \setminus T \times V_H) \cup \{ (t,v) \mid T \times \{v\} \cap E_H \neq \emptyset \}$. Regarding the root hyperedge $s_{H/T}$, if $s_H \in T$ then $s_{H/T} = t$, otherwise it remains the same s_H.

Example 2. *Considering two vertices $v, w \in V_H$, we denote by $H_{v=w}$ the contraction on vertices in $H/\{v,w\}$. For two hyperedges $s, t \in S_H$, we denote by $H_{s=t}$ the contraction on hyperedges in $H/\{s,t\}$. For the hypergraph H used in the previous example, the contraction on vertices $H_{w=w'}$ and the contraction on hyperedges $H_{t=t'}$ are depicted in Figure 2.*

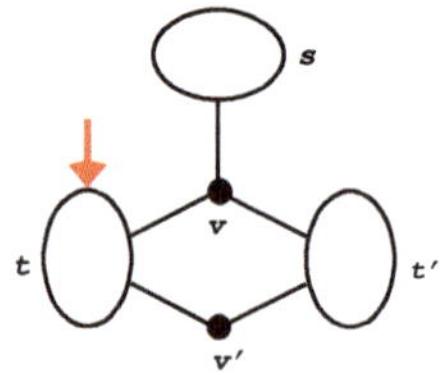

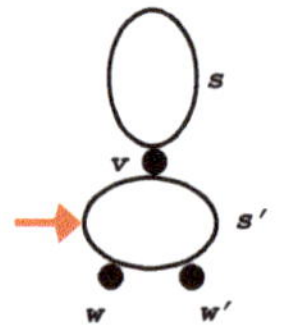

Figure 2. A contraction on vertices (**left**), and a contraction on hyperedges (**right**).

The *isomorphism* between two hypergraphs H and H' is defined by two bijections $\phi_S : S_H \to S_{H'}$ and $\phi_V : V_H \to V_{H'}$ which satisfy $\phi_S(s_H) = s_{H'}$ and $(s,v) \in E_H$ if and only if $(\phi_S(s), \phi_V(v)) \in E_{H'}$ for all $s \in S_H$ and $v \in V_H$. In such a situation, we say that hypergraphs H and H' are isomorphic (denoted by $H = H'$).

The isomorphism relation is an equivalence over hypergraphs. The names of hyperedges and vertices do not play any role in the isomorphism of hypergraphs. For the graphical representation of an equivalence class, it can be used any hypergraph (after removing the names of hyperedges and vertices).

4. Control Structure for Pattern Calculus

Milner proposed the control structures and action calculi as a unifying framework for the models of concurrent systems in [8]. A control structure defines the static aspects of a process calculus, while the corresponding action calculus describes various models of interactive behaviours. Regarding the behaviour, distinct action calculi differ only in their generators (called controls). Thus, the previously mentioned Petri nets and π-nets, as well as our hypergraph model, differ only in their generators. Analyzing these generators, it is possible to compare and classify the formal models for concurrent systems. Moreover, by selecting some specific generators, it is possible to combine existing models in order to obtain a new desired model.

The control structures presented here follow the definitions of [9]. Essentially, a control structure is defined by a set of terms, an equational theory, and a reduction relation over terms. This fact is condensed in the following expression:

$$ControlStructure = Actions + EquationalTheory + Reaction.$$

From an algebraic viewpoint, a control structure is a symmetric strict monoidal category with an additional structure [10]. The morphisms of the symmetric strict monoidal category correspond to the terms of the control structure; they are denoted by $a, b, c, \ldots$ and called *actions*.

The control structure uses an enumerable set $\mathcal{X}$ of *names* together with a signature $(P, \mathcal{K})$ in which P is a set of prime arities and $\mathcal{K}$ is a set of control operators. Every name $x \in \mathcal{X}$ has a prime arity $p \in P$, and this is denoted by $x : p$. Each control $K \in \mathcal{K}$ has an arity rule. In addition to the specific control operators, every control structure contains a *datum* operator $\langle x \rangle : \epsilon \to p$ (where $x : p$) and a *discard* operator $\omega_p : p \to \epsilon$, as well as an *abstractor* operator $\mathtt{ab}_x\, a : p \otimes m \to p \otimes n$ (where $x : p$ and $a : m \to n$). The equational theory is the same for all control structures. To express the evolution, a set R of *reaction rules* is used; reaction rules are ordered pairs of terms with the same arity.

Each action a possesses a *surface* $\mathtt{surf}(a) = \{x \in X \mid \exists p.x : p \text{ and } \mathtt{ab}_x\, a \neq \mathtt{id}_p \otimes a\}$.

The equality $=$ between actions is valid whenever the equation $a = b$ could be proved by using the axioms of the control structure; otherwise $a \neq b$.

We present here some results used later in the proofs of our results.

Proposition 2. *The following properties hold in any control structure:*

$$\mathtt{surf}(\langle x \rangle) \subseteq \{x\} \quad \mathtt{surf}(\omega) = \emptyset \quad \mathtt{surf}(a \otimes b) \subseteq \mathtt{surf}(a) \cup \mathtt{surf}(b).$$

The following properties hold in any control structure, whenever $x \notin \mathtt{surf}(c)$*:*

$$\mathtt{surf}(\mathtt{id}) = \emptyset \quad \mathtt{surf}(\mathtt{p}) = \emptyset \quad \mathtt{surf}(a \cdot b) \subseteq \mathtt{surf}(a) \cup \mathtt{surf}(b).$$

Additionally,

1. $\mathtt{p}_{n,\epsilon} = \mathtt{id}_n$
2. $a \otimes b = b \otimes a$ $\quad (a, b : \epsilon \to \epsilon)$
3. $(x)(c \cdot b) = (\mathtt{id}_p \otimes c) \cdot (x)b$ $\quad (x : p)$
4. $(x)(c \otimes b) = c \otimes (x)b$ $\quad (c : \epsilon \to n)$
5. $(x)(a \otimes c) = (x)a \otimes c$
6. $(z)(y)a = (\mathtt{p}_{p,q} \otimes \mathtt{id}_m) \cdot (y)(z)a$ $\quad (z : p, y : q, a : m \to n)$

Proposition 3. *The following properties are provable in any control structure:*

1. $[x/y](a \cdot b) = [x/y]a \cdot [x/y]b$;
2. $[x/y](a \otimes b) = [x/y]a \otimes [x/y]b$;
3. $[x/y](z)a = (z)[x/y]a$ $\quad$ *if* $z \notin \{x, y\}$;
4. $[x/y](x)a = (w)[x/y][w/x]a$ $\quad$ *if* $w \notin \mathtt{surf}(a) \cup \{x, y\}$ *and* $x \neq y$.

We define the control structure for our pattern calculus, emphasizing on its actions. We also present a graphical representation for its processes. The monoid $(\mathbb{N}, +, 0)$ of the natural numbers provides the arity monoid of the control structure, with $m, n, k, \ldots$ ranging over natural numbers, and $[n] = \{1, 2, \ldots, n\}$ denoting the first n natural numbers. The (unique) prime arity 1 is associated with each name $x \in \mathcal{X}$. For a number k and a function $f : [n] \to Y$, we define $k \oplus f : \{k+1, \ldots k+n\} \to Y$ by $(k \oplus f)(i) = f(i-k)$. Following [9], the control structure is defined over the set $\mathcal{X} = \{z_i \mid i \in \mathbb{N}\}$ of names using $x, y, u, \ldots$ as meta-variables.

Regarding the actions of the control structure for our calculus, they are given by enriched hypergraphs called shortly *pattern nets*. An action $a = (H, \Sigma)$ with arity $m \to n$ is given by a hypergraph H together with its decoration $\Sigma = \langle \mathtt{I}, \mathtt{O}, \lambda, \tau, \mu \rangle$ consisting of:

- An *input function* given by an injective function $\mathtt{I} : [m] \to V_H$;
- An *output function* given by a function $\mathtt{O} : [n] \to V_H$;
- A *label function* given by an injective function $\lambda : Z \to V_H$, where $Z \subseteq \mathcal{X}$;
- A *transition relation* $\tau \subseteq V_H \times V_H$;
- A *resource function* $\mu : S_H \to \mathbb{N}^{V_H \times V_H}$.

We can look at these functions as multisets over $S_H \times V_H \times V_H$. We denote by $\{x, y, y\}$ a multiset μ over $\{x, y, z\}$ such that $\mu(x) = 1$, $\mu(y) = 2$ and $\mu(z) = 0$; we use the standard multiset operations over these functions: $(\uplus, -, \ldots)$.

We extend in a straightforward way the isomorphism and contraction introduced for hypergraphs. Considering $a_i = (H_i, \Sigma_i)$ with $\Sigma_i = \langle \mathtt{I}_i, \mathtt{O}_i, \lambda_i, \tau_i, \mu_i \rangle$ $(i \in [2])$, the nets a_1 and a_2 are isomorphic if there is a hypergraph isomorphism (ϕ_S, ϕ_V) between H_1 and H_2 such that $\phi_V \circ \mathtt{I}_1 = \mathtt{I}_2$, $\phi_V \circ \mathtt{O}_1 = \mathtt{O}_2$, $\phi_V \circ \lambda_1 = \lambda_2$, and $(v, v') \in \tau_1$ if and only if

$(\phi_V(v), \phi_V(v')) \in \tau_2$, together with $\mu_1(s, v, v') = \mu_2(\phi_S(s), \phi_V(v), \phi_V(v'))$ for all $s \in S_{H_1}$ and $v, v' \in V_{H_1}$.

For the graphical representations of the pattern nets, let us consider a generic net $a = (H, \Sigma)$ with $\Sigma = \langle \mathtt{I}, \mathtt{O}, \lambda, \tau, \mu \rangle$; the hypergraph H is presented by assuming that its lines are of length zero (see Figure 1):

- Whenever $\mathtt{I}(i) = v$, $\mathtt{O}(k) = v'$ and $\lambda(x) = w$, an *input label* (i) is assigned to vertex v, an *output label* $\langle k \rangle$ to vertex v', and a *name label* x to vertex w;
- Whenever $(v, v') \in \tau$, an arc is drawn outside any oval from vertex v to vertex v';
- Whenever $\mu(s, v, v') > 0$, v and v' lie on the same hyperedge s: $(s, v), (s, v') \in E_H$; more exactly, whenever $\mu(s, v, v') = k > 0$, we have k arcs inside the oval s from vertex v to vertex v'.

As for hypergraphs, isomorphic nets are not distinguished. The names of vertices and hyperedges do not play any role in the isomorphic nets, and so the graphical representation of an isomorphic (equivalence) class of pattern nets is given by any net of the class after removing the names of vertices and hyperedges.

The *control structure operators* for our pattern nets are:

- *datum* $\langle x \rangle^\gamma = (H, \Sigma) : 0 \to 1$ defined by
$H = \langle \{s\}, \{v\}, \{(s, v)\}, s \rangle$ and
$\Sigma = \langle \emptyset, \{1 \mapsto v\}, \{x \mapsto v\}, \emptyset, \emptyset \rangle$

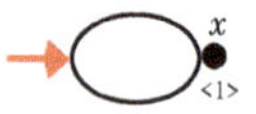

- *discard* $\omega^\gamma = (H, \Sigma) : 1 \to 0$ defined by
$H = \langle \{s\}, \{v\}, \{(s, v)\}, s \rangle$ and
$\Sigma = \langle \{1 \mapsto v\}, \emptyset, \emptyset, \emptyset, \emptyset \rangle$

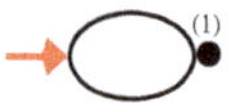

The three *controls* generating the pattern nets are:

- $\nu^\gamma = (H, \Sigma) : 0 \to 1$ defined by
$H = \langle \{s\}, \{v\}, \{(s, v)\}, s \rangle$ and
$\Sigma = \langle \emptyset, \{1 \mapsto v\}, \emptyset, \emptyset, \emptyset \rangle$

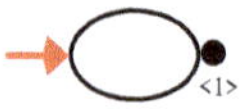

- $\mathtt{out}^\gamma = (H, \Sigma) : 2 \to 0$ defined by
$H = \langle \{s\}, \{v, v'\}, \{(s, v), (s, v')\}, s \rangle$ and
$\Sigma = \langle \emptyset, \{1 \mapsto v, 2 \mapsto v'\}, \emptyset, \emptyset, \{(s, v', v)\} \rangle$

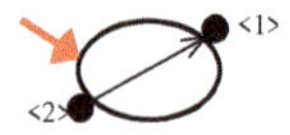

- If $a = (H, \Sigma) : 1 \to 0$ and $\Sigma = \langle \mathtt{I}, \mathtt{O}, \lambda, \tau, \mu \rangle$, then
$\mathtt{def}^\gamma\, a = (H', \Sigma') : 1 \to 0$, where
$H' = \langle S_H \cup \{t\}, V_H \cup \{v\}, E_H \cup \{(t, v)\}, t \rangle$, for fresh $t \notin S_H$ and $v \notin V_H$, and
$\Sigma' = \langle \{1 \mapsto v\}, \mathtt{O}, \lambda, \tau \cup \{(v, \mathtt{I}(1))\}, \mu \rangle$.

The *equational theory* is defined by the following operators.

Let us consider the nets $a_i = (H_i, \Sigma_i)$ with $\Sigma_i = \langle \mathtt{I}_i, \mathtt{O}_i, \lambda_i, \tau_i, \mu_i \rangle$ and $\lambda_i : Z_i \to V_{H_i}$, where $i \in [2]$. Without loss of generality, we consider $s_{H_1} = s_{H_2} = s$. $(S_{H_1} - \{s_{H_1}\}) \cap (S_{H_2} - \{s_{H_2}\}) = \emptyset$ and $\lambda_1(z) = \lambda_2(z)\ \forall z \in Z_1 \cap Z_2$, as well as $(V_{H_1} - \lambda_1(Z_1 \cap Z_2)) \cap (V_{H_2} - \lambda_2(Z_1 \cap Z_2)) = \emptyset$.

- *Identity* $\mathtt{id}_m^\gamma = (H, \Sigma) : m \to m$ defined by
$H = \langle \{s\}, \{v_i | i \in [m]\}, \{(s, v_i) | i \in [m]\}, s \rangle$ and
$\Sigma = \langle \{i \mapsto v_i | i \in [m]\}, \{i \mapsto v_i | i \in [m]\}, \emptyset, \emptyset, \emptyset \rangle$

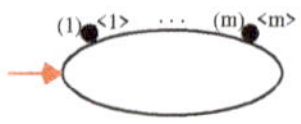

- *Symmetry* $\mathtt{p}_{m,n}^\gamma = (H, \Sigma) : m + n \to n + m$ defined by
$H = \langle \{s\}, \{v_i | i \in [m+n]\}, \{(s, v_i) | i \in [m+n]\}, s \rangle$
$\Sigma = \langle \{i \mapsto v_i | i \in [m+n]\}, \{i \mapsto v_{m+i} | i \in [n]\} \cup \{n + i \mapsto v_i | i \in [m]\}, \emptyset, \emptyset, \emptyset \rangle$

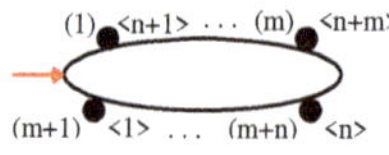

- *Tensorial product* $a_1 \otimes a_2 : m + k \to n + l$ of two nets $a_1 : m \to n$ and $a_2 : k \to l$ is obtained by combining them as follows: in a_2, the input labels are incremented by m and the output labels are incremented by n; then overlap the two roots and the vertices of a_1 and of a_2 with the same name. Formally, $a_1 \otimes a_2 = (H, \Sigma)$, where

$$H = \langle S_{H_1} \cup S_{H_2}, V_{H_1} \cup V_{H_2}, E_{H_1} \cup E_{H_2}, s \rangle \text{ and}$$

$$\Sigma = \langle \mathtt{I}_1 \cup m \oplus \mathtt{I}_2, \mathtt{O}_1 \cup n \oplus \mathtt{O}_2, \lambda_1 \cup \lambda_2, \tau_1 \cup \tau_2, \mu_1 \uplus \mu_2 \rangle.$$

– *Composition* $a_1 \cdot a_2 : m \to k$ of two nets $a_1 : m \to n$ and $a_2 : n \to k$ is obtained by combining them as follows: overlap the two roots and vertices of a_1 and a_2 with the same name; for every $i \in [n]$, overlap the vertex labelled $\langle i \rangle$ in a_1 with the vertex labelled (i) in a_2, and then remove the labels (i) and $\langle i \rangle$.

Formally, $a_1 \cdot a_2 = (H, \Sigma)_{\mathtt{O}_1(1)=\mathtt{I}_2(1),\ldots,\mathtt{O}_1(n)=\mathtt{I}_2(n)}$, where

$$H = \langle S_{H_1} \cup S_{H_2}, V_{H_1} \cup V_{H_2}, E_{H_1} \cup E_{H_2}, s \rangle \text{ and}$$

$$\Sigma = \langle \mathtt{I}_1, \mathtt{O}_2, \lambda_1 \cup \lambda_2, \tau_1 \cup \tau_2, \mu_1 \uplus \mu_2 \rangle.$$

– *Abstractor.* Let us consider a net $a = (H, \Sigma) : m \to n$ with $\Sigma = \langle \mathtt{I}, \mathtt{O}, \lambda, \tau, \mu \rangle$.

Then $\mathtt{ab}^\gamma_x\, a : 1 + m \to 1 + n$ is obtained from a in the following steps: increment all the input and output labels by 1; assign both the input label (1) and the output label $\langle 1 \rangle$ to vertex x, and then remove the label x. Formally, $\mathtt{ab}^\gamma_x\, a = (H, \Sigma')$, where

$$\Sigma' = \langle \{1 \mapsto \lambda(x)\} \cup 1 \oplus \mathtt{I}, \{1 \mapsto \lambda(x)\} \cup 1 \oplus \mathtt{O}, \lambda - \{x \mapsto \lambda(x)\}, \tau, \mu \rangle.$$

In general, these operators over the pattern nets are well-defined. However, the abstractor $\mathtt{ab}^\gamma_x\, a$ is not well-defined if a vertex labelled by x is not contained in the net a. To avoid such a situation, we adjust the definition of the above operators by

$$\mathtt{op}(a, \ldots) \stackrel{def}{=} \mathtt{op}^\gamma(a \otimes^\gamma \mathbf{i}, \ldots) \otimes^\gamma \mathbf{i},$$

where op stands for each operator defined above, and $\mathbf{i} = (H, \Sigma)$ is the pattern net

$$H = \langle \{s\}, \{v_i | i \in \mathbb{N}\}, \{(s, v_i) | i \in \mathbb{N}\}, s \rangle$$

$$\Sigma = \langle \emptyset, \emptyset, \{z_i \mapsto v_i | i \in \mathbb{N}\}, \emptyset, \emptyset \rangle.$$

Following [9], it is not difficult to prove the following result.

Proposition 4. *The operators* $\langle x \rangle$, ω, ν, $\mathtt{out}$, $\mathtt{def}$, $\mathtt{id}$, $\mathtt{p}$, $\cdot$, $\otimes$ *and* $\mathtt{ab}_x$ *define a control structure.*

The actions of this control structure determine the *hypergraph model* for the pattern calculus. We actually use the derived control operators:

$$\mathtt{out}_u \stackrel{def}{=} (\langle u \rangle \otimes \mathtt{id}_1) \cdot \mathtt{out};$$
$$\mathtt{def}_u\, a \stackrel{def}{=} \langle u \rangle \cdot \mathtt{def}\; a.$$

The reaction $\searrow$ is the smallest relation over the pattern nets closed under equality, composition, tensorial product and abstraction which satisfies the control rule

$$\mathtt{out}_u \otimes \mathtt{def}_u\, a \searrow a \otimes \mathtt{def}_u\, a.$$

The corresponding graphical description of the reaction rule is given by:

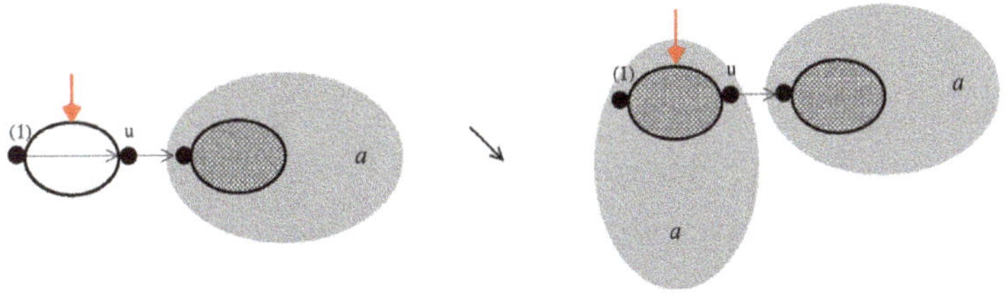

In this diagram, the scope of the def operator is represented as a gray patch; due to the properties derived from the syntax of our calculus, this patch can actually be determined from the hypergraph structure.

The operators, actions and reaction complete the definition of our nets.

It is worth noting that the def operator can be generalized, namely we can have a more general control operator $\mathtt{def}_{u_1 \ldots u_m}$ by $\mathtt{def}_{u_1 \ldots u_m}\, a = (\langle u_1 \rangle \otimes \ldots \otimes \langle u_m \rangle) \cdot \mathtt{def}\ a$. Moreover, the corresponding graphical representation is extended by using m external arcs to connect the new root hyperedge to the old one. The corresponding reaction is generalized in the following way: $\mathtt{out}_{u_1} \otimes \ldots \otimes \mathtt{out}_{u_m} \otimes \mathtt{def}_{u_1 \ldots u_m}\, a \searrow a \otimes \mathtt{def}_{u_1 \ldots u_m}\, a$.

We present some proprieties of the pattern nets. The proofs of these properties are tedious (but easy), based mainly on definitions and the structure of the nets.

Lemma 2. *We have the following properties:*

1. $\mathtt{surf}(\mathtt{out}_u) \subseteq \{u\}$ *and* $\mathtt{surf}(\mathtt{def}_u\, a) \subseteq \{u\} \cup \mathtt{surf}(a)$.
2. *For any substitution* $\sigma = \{x/y\}$, $[x/y]\mathtt{out}_u = \mathtt{out}_{\sigma u}$ *and* $[x/y]\mathtt{def}_u\, a = \mathtt{def}_{\sigma u}\, [x/y]a$.
3. *If* $a \searrow b$, *then there exists* b' *such that* $b = b'$ *and* $\mathtt{surf}(b') \subseteq \mathtt{surf}(a)$.
4. $\langle v \rangle \cdot \mathtt{out}_w \otimes \mathtt{def}_u\, (y)a \searrow b$ *iff* $u = w$ *and* $b = [v/y]a \otimes \mathtt{def}_u\, (y)a$.
5. $a_1 \otimes a_2 \otimes a_3 \searrow c$ *iff*
   - *either there exists* $i \in [3]$ *such that* $a_i \searrow b$ *and* $c = b \otimes a_j \otimes a_k$, *or*
   - *there exist* $i, j \in [3]$ *such that* $a_i \otimes a_j \searrow b$ *and* $c = b \otimes a_k$, *where* $[3] = \{i, j, k\}$.
6. *Whenever* $u \notin \mathtt{surf}(b)$, $b \otimes \mathtt{def}_u\, a \searrow c$ *iff* $b \searrow b'$ *and* $c = b' \otimes \mathtt{def}_u\, a$.
7. $\nu \cdot (x)a \searrow b$ *iff* $a \searrow a'$ *and* $b = \nu \cdot (x)a'$.

5. Fully Abstract Hypergraph Model of the Pattern Calculus

This section presents the main results of the paper. These results reveal the hypergraphs as a fully abstract model for the pattern calculus. According to [11], a model is fully abstract if all observationally equivalent terms in the object language represent the same object in the model. This means that processes with different behaviour are not mapped to the same hypergraph. Moreover, we prove a correspondence between the reduction of the processes and the reduction of their hypergraph representation.

Definition 4. *The semantic relationship* $[\![-]\!]$ *between the pattern calculus processes and the pattern nets is defined by structural induction as follows:*

1. $[\![0]\!] = \mathtt{id}_0\,;$
2. $[\![u\langle v \rangle]\!] = \langle v \rangle \cdot \mathtt{out}_u\,;$
3. $[\![P \mid Q]\!] = [\![P]\!] \otimes [\![Q]\!]\,;$
4. $[\![\mathtt{def}\ u\langle y \rangle \triangleright P\ \mathtt{in}\ Q]\!] = \nu \cdot (u)(\, [\![Q]\!] \otimes \mathtt{def}_u\, (y)\, [\![P]\!]\,)\,.$

We prove some results involving this semantic relationship $[\![-]\!]$.

Lemma 3. *For every process* $P \in \mathcal{P}$, *we have* $[\![P]\!] : 0 \to 0$.

Proof. A simple induction on the structure of P. In the case of our nets, 0 is the neutral element of the arity monoid $(\mathbb{N}, +, 0)$. For case (4) of the previous definition, we use the discard operator ω instead of ω_p. Since 1 is the only prime arity p of the monoid $(\mathbb{N}, +, 0)$, we omit the index without any risk of confusion. □

Lemma 4. *For every process* $P \in \mathcal{P}$, *we have* $\mathtt{surf}(\, [\![P]\!]\,) \subseteq \mathtt{fn}(P)$.

Proof. By induction on the structure of P (the proof uses Lemma 2). □

Lemma 5. *For two names* $x, y \in X$ *and a process* $P \in \mathcal{P}$, *we have* $[\![\{x/y\}P]\!] = [x/y]\, [\![P]\!]$.

Proof. Induction on the definition of the substitution over processes (and Lemma 5). □

Proposition 5. *If* $P \equiv Q$, *then* $[\![P]\!] = [\![Q]\!]$.

Proof. Induction on the definition of structural congruence. Let us consider the relation

$$\sim = \{(P,Q) \in \mathcal{P} \mid P \equiv Q \text{ and } [\![P]\!] = [\![Q]\!] \}.$$

Proof is reduced to the equality between $\sim$ and $\equiv$. Obviously, $\sim \subseteq \equiv$. We show that $\sim$ satisfies the axioms from the definition of $\equiv$. Since $\equiv$ is the smallest relation satisfying these axioms, it follows that $\equiv \subseteq \sim$, and so $\sim = \equiv$. Thus, to prove that $\equiv \subseteq \sim$, it is enough to verify that $\sim$ satisfies the axioms from the definition of $\equiv$.

The cases $P \mid 0 \equiv P$, $P \mid Q \equiv Q \mid P$ and $(P \mid Q) \mid R \equiv P \mid (Q \mid R)$ are rather trivial, based on the fact that $\mathtt{id}_0$ is neutral for tensor product, together with the commutativity and associativity of tensor product $\otimes$ in the equational theory (of the control structures).

Let us consider the other cases.

– $\mathtt{def}\ u\langle v\rangle \triangleright P\ \mathtt{in}\ Q \equiv \mathtt{def}\ u\langle t\rangle \triangleright \{t/v\}P\ \mathtt{in}\ Q$, if $t \notin \mathtt{fn}(P)$.
Assume $t \notin \mathtt{fn}(P)$. By Lemma 4, it follows that $t \notin \mathtt{surf}([\![P]\!])$. Then,

$$\begin{aligned}
&[\![\mathtt{def}\ u\langle t\rangle \triangleright \{t/v\}P\ \mathtt{in}\ Q]\!] = \\
&= \nu \cdot (u)([\![Q]\!] \otimes \mathtt{def}_u\,(t)\,[\![\{t/v\}P]\!]) && \text{by Lemma 5} \\
&= \nu \cdot (u)([\![Q]\!] \otimes \mathtt{def}_u\,(t)[t/v]\,[\![P]\!]) \\
&= [\![\mathtt{def}\ u\langle v\rangle \triangleright P\ \mathtt{in}\ Q]\!].
\end{aligned}$$

– $\mathtt{def}\ u\langle v\rangle \triangleright P\ \mathtt{in}\ Q \equiv \mathtt{def}\ w\langle v\rangle \triangleright \{w/u\}P\ \mathtt{in}\ \{w/u\}Q$ if $v \notin \{u,w\}$, $w \notin \mathtt{fn}(P \mid Q)$.
Assume $v \notin \{u,w\}$ and $w \notin \mathtt{fn}(P \mid Q)$; then, $u \neq v$ and $w \notin \mathtt{fn}(P) \cup \mathtt{fn}(Q) \cup \{v\}$. By Lemma 4, $w \notin \mathtt{surf}([\![P]\!]) \cup \mathtt{surf}([\![Q]\!])$. If $u = w$, then the result is trivial.
Let us assume that $u \neq w$.

$$\begin{aligned}
&[\![\mathtt{def}\ w\langle v\rangle \triangleright \{w/u\}P\ \mathtt{in}\ \{w/u\}Q]\!] = \\
&= \nu \cdot (w)([\![\{w/u\}Q]\!] \otimes \mathtt{def}_w\,(v)\,[\![\{w/u\}P]\!]) && \text{by Lemma 5} \\
&= \nu \cdot (w)([w/u]\,[\![Q]\!] \otimes \mathtt{def}_w\,(v)[w/u]\,[\![P]\!]) && \text{by Proposition 3 and Lemma 2} \\
&= \nu \cdot (w)[w/u]([\![Q]\!] \otimes \mathtt{def}_u\,(v)\,[\![P]\!]) && \text{by Lemma 2} \\
&= [\![\mathtt{def}\ u\langle v\rangle \triangleright P\ \mathtt{in}\ Q]\!].
\end{aligned}$$

– $Q_1 \mid \mathtt{def}\ u\langle v\rangle \triangleright P\ \mathtt{in}\ Q_2 \equiv \mathtt{def}\ u\langle v\rangle \triangleright P\ \mathtt{in}\ (Q_1 \mid Q_2)$ if $u \notin \mathtt{fn}(Q_1)$.
Assume $u \notin \mathtt{fn}(Q_1)$. By Lemma 4, $u \notin \mathtt{surf}([\![Q_1]\!])$. Then,

$$\begin{aligned}
&[\![\mathtt{def}\ u\langle v\rangle \triangleright P\ \mathtt{in}\ (Q_1 \mid Q_2)]\!] = \\
&= \nu \cdot (u)([\![Q_1]\!] \otimes [\![Q_2]\!] \otimes \mathtt{def}_u\,(v)\,[\![P]\!]) && \text{by Lemma 3 and Proposition 2} \\
&= [\![Q_1 \mid \mathtt{def}\ u\langle v\rangle \triangleright P\ \mathtt{in}\ Q_2]\!].
\end{aligned}$$

– $\mathtt{def}\ u\langle v\rangle \triangleright P_1\ \mathtt{in}\ \mathtt{def}\ w\langle t\rangle \triangleright P_2\ \mathtt{in}\ Q \equiv \mathtt{def}\ w\langle t\rangle \triangleright P_2\ \mathtt{in}\ \mathtt{def}\ u\langle v\rangle \triangleright P_1\ \mathtt{in}\ Q$
if $u \neq w$, $u \notin \mathtt{fn}(P_2)$, and $w \notin \mathtt{fn}(P_1)$.
Assume $u \neq w$, $u \notin \mathtt{fn}(P_2)$ and $w \notin \mathtt{fn}(P_1)$. By Lemma 4, it follows that $u \notin \mathtt{surf}([\![P_2]\!])$ and $w \notin \mathtt{surf}([\![P_1]\!])$. Furthermore, by Lemma 2, $w \notin \mathtt{surf}(\mathtt{def}_u\,(v)\,[\![P_1]\!]$.

$$\begin{aligned}
&[\![\mathtt{def}\ u\langle v\rangle \triangleright P_1\ \mathtt{in}\ \mathtt{def}\ w\langle t\rangle \triangleright P_2\ \mathtt{in}\ Q]\!] = \\
&= \nu \cdot (u)(\nu \cdot (w)([\![Q]\!] \otimes \mathtt{def}_w\,(t)\,[\![P_2]\!]) \otimes \mathtt{def}_u\,(v)\,[\![P_1]\!]) && \text{by Proposition 2} \\
&= \nu \cdot (u)(\nu \cdot (w)([\![Q]\!] \otimes \mathtt{def}_w\,(t)\,[\![P_2]\!] \otimes \mathtt{def}_u\,(v)\,[\![P_1]\!])) && \text{by Proposition 2} \\
&= (\nu \otimes \nu) \cdot (u)(w)([\![Q]\!] \otimes \mathtt{def}_w\,(t)\,[\![P_2]\!] \otimes \mathtt{def}_u\,(v)\,[\![P_1]\!]) && = \mathbf{X}.
\end{aligned}$$

In a similar way, we obtain $[\![\mathtt{def}\ w\langle t\rangle \triangleright P_2\ \mathtt{in}\ \mathtt{def}\ u\langle v\rangle \triangleright P_1\ \mathtt{in}\ Q]\!] =$

$$= (\nu \otimes \nu) \cdot (w)(u)([\![Q]\!] \otimes \mathtt{def}_u\,(v)\,[\![P_1]\!] \otimes \mathtt{def}_w\,(t)\,[\![P_2]\!]) \quad = \mathbf{Y}.$$

To complete the proof, it remains to prove that $\mathbf{X} = \mathbf{Y}$.

$$\begin{aligned}
&\mathbf{X} = && \text{by Proposition 2} \\
&= (\nu \otimes \nu) \cdot \mathtt{p}_{1,1} \cdot (w)(u)([\![Q]\!] \otimes \mathtt{def}_w\,(t)\,[\![P_2]\!] \otimes \mathtt{def}_u\,(v)\,[\![P_1]\!]) \\
&= \mathtt{p}_{0,0} \cdot (\nu \otimes \nu) \cdot (w)(u)([\![Q]\!] \otimes \mathtt{def}_u\,(v)\,[\![P_1]\!] \otimes \mathtt{def}_w\,(t)\,[\![P_2]\!]) \\
&= \mathbf{Y}.
\end{aligned}$$

□

Theorem 1. *If* $P \to Q$, *then* $[\![P]\!] \searrow [\![Q]\!]$.

Proof. By induction on the definition of $P \to Q$.
* r1: $P \to Q$ is

$$\texttt{def}\ u_1\langle y_1\rangle \triangleright Q_1\ \texttt{in def}\ u_2\langle y_2\rangle \triangleright Q_2\ \texttt{in}\ \ldots\ \texttt{def}\ u_n\langle y_n\rangle \triangleright Q_n\ \texttt{in}\ R \mid u_i\langle v\rangle \ \to$$

$$\texttt{def}\ u_1\langle y_1\rangle \triangleright Q_1\ \texttt{in def}\ u_2\langle y_2\rangle \triangleright Q_2\ \texttt{in}\ \ldots\ \texttt{def}\ u_n\langle y_n\rangle \triangleright Q_n\ \texttt{in}\ R \mid \{v/y_i\}Q_i,$$ where

$\{u_{i+1},\ldots,u_n\} \cap (\texttt{fn}(Q_i) \cup \{u_i\}) = \emptyset$, $i \in [n]$, and $n \geq 1$. According to Lemma 2 and Lemma 4, $\{u_{i+1},\ldots,u_n\} \cap \texttt{surf}(\texttt{def}_{u_i}(y_i) [\![Q_i]\!]) = \emptyset$. According to Proposition 2 and using the compatibility of $\searrow$ with composition, tensorial product and abstraction,
$[\![P]\!] = \nu \cdot (u_1)(\texttt{def}_{u_1}(y_1) [\![Q_1]\!] \otimes$

$$\begin{aligned}
= \quad & \vdots \\
& \nu \cdot (u_i)(\texttt{def}_{u_i}(y_i) [\![Q_i]\!] \otimes \\
& \quad \vdots \\
& \quad \nu \cdot (u_n)(\texttt{def}_{u_n}(y_n) [\![Q_n]\!] \otimes [\![R]\!] \otimes \langle v\rangle \cdot \texttt{out}_{u_i})\ldots)\ldots) \\
= \quad & \nu \cdot (u_1)(\texttt{def}_{u_1}(y_1) [\![Q_1]\!] \otimes \qquad \text{by Lemma 2} \\
& \quad \vdots \\
& \nu \cdot (u_{i-1})(\texttt{def}_{u_{i-1}}(y_{i-1}) [\![Q_{i-1}]\!] \otimes \\
& \nu \cdot (u_i)(\\
& \nu \cdot (u_{i+1})(\texttt{def}_{u_{i+1}}(y_{i+1}) [\![Q_{i+1}]\!] \otimes \\
& \quad \vdots \\
& \quad \nu \cdot (u_n)(\texttt{def}_{u_n}(y_n) [\![Q_n]\!] \otimes [\![R]\!] \otimes \langle v\rangle \cdot \texttt{out}_{u_i} \otimes \texttt{def}_{u_i}(y_i) [\![Q_i]\!])\ldots)))\ldots) \\
\searrow \quad & \nu \cdot (u_1)(\texttt{def}_{u_1}(y_1) [\![Q_1]\!] \otimes \qquad \text{by Lemma 5} \\
& \quad \vdots \\
& \nu \cdot (u_i)(\texttt{def}_{u_i}(y_i) [\![Q_i]\!] \otimes \\
& \quad \vdots \\
& \quad \nu \cdot (u_n)(\texttt{def}_{u_n}(y_n) [\![Q_n]\!] \otimes [\![R]\!] \otimes [v/y_i] [\![Q_i]\!])\ldots)\ldots) \\
= \quad & [\![Q]\!].
\end{aligned}$$

* r2: $P \to Q$ is $\texttt{def}\ u\langle v\rangle \triangleright R\ \texttt{in}\ P' \to \texttt{def}\ u\langle v\rangle \triangleright R\ \texttt{in}\ Q'$ with $P' \to Q'$.

By induction, $[\![P']\!] \searrow [\![Q']\!]$. Since $\searrow$ is closed under composition, tensor and abstraction, it follows that

$$[\![P]\!] = \nu \cdot (u)([\![P']\!] \otimes \texttt{def}_u(v) [\![R]\!]) \searrow \nu \cdot (u)([\![Q']\!] \otimes \texttt{def}_u(v) [\![R]\!]) = [\![Q]\!].$$

* r3: $P \to Q$ with $P \equiv P'$, $P' \to Q'$ and $Q' \equiv Q$.

By the induction hypothesis, $[\![P']\!] \searrow [\![Q']\!]$. By Proposition 5, we have $[\![P]\!] = [\![P']\!]$ and $[\![Q']\!] = [\![Q]\!]$. Since $\searrow$ is closed under equality, then $[\![P]\!] \searrow [\![Q]\!]$. □

Lemma 6. *$\langle v\rangle \cdot \texttt{out}_u \otimes [\![P]\!] \searrow a$ iff $[\![P]\!] \searrow b$ and $a = \langle v\rangle \cdot \texttt{out}_u \otimes b$.*

Proof. ($\Leftarrow$) A consequence of the fact that the reaction is closed under tensorial product and equality.
($\Rightarrow$) Induction on the structure of P.

- If P is the empty process or a message, then $\langle v\rangle \cdot \texttt{out}_u \otimes [\![P]\!] \not\searrow$. Therefore, the statement of the lemma is obviously true because its premise is not satisfied.
- If P is a parallel composition $P_1 \mid P_2$, then $\langle v\rangle \cdot \texttt{out}_u \otimes [\![P_1]\!] \otimes [\![P_2]\!] \searrow a$. Since $\langle v\rangle \cdot \texttt{out}_u \not\searrow$, it follows (Lemma 2) that one of the following cases remains possible:

(1) $[\![P_i]\!] \searrow b'$ and $a = \langle v\rangle \cdot \texttt{out}_u \otimes b' \otimes [\![P_j]\!]$;
(2) $[\![P_i]\!] \otimes [\![P_j]\!] \searrow b'$ and $a = \langle v\rangle \cdot \texttt{out}_u \otimes b'$;

(3) $\langle v\rangle \cdot \mathtt{out}_u \otimes [\![P_i]\!] \searrow a'$ and $a = a' \otimes [\![P_j]\!]$, where $\{i,j\} = [2]$.

Note that by Proposition 2, we have $[\![P]\!] = [\![P_i]\!] \otimes [\![P_j]\!]$.
In case (1), $[\![P]\!] \searrow b' \otimes [\![P_j]\!]$, and we consider $b = b' \otimes [\![P_j]\!]$. In case (2), we take $b = b'$. In case (3), by induction, we have $[\![P_i]\!] \searrow b'$ and $a' = \langle v\rangle \cdot \mathtt{out}_u \otimes b'$. Thus, $[\![P]\!] \searrow b' \otimes [\![P_j]\!]$, considering $b = b' \otimes [\![P_j]\!]$.

- If P is a definition $\mathtt{def}\ w\langle t\rangle \triangleright P_1\ \mathtt{in}\ P_2$, then we may assume without losing generality that $w \notin \{u,v\}$. It follows from Lemma 2 together with Proposition 2 that $\nu \cdot (w)(\langle v\rangle \cdot \mathtt{out}_u \otimes [\![P_2]\!] \otimes \mathtt{def}_w(t)\, [\![P_1]\!]) \searrow a$. By Lemma 2, $\langle v\rangle \cdot \mathtt{out}_u \otimes [\![P_2]\!] \otimes \mathtt{def}_w(t)\, [\![P_1]\!] \searrow a'$ and $a = \nu \cdot (w)a'$. Since $\langle v\rangle \cdot \mathtt{out}_u \not\searrow$, $\mathtt{def}_w(t)\, [\![P_1]\!] \not\searrow$ and $\langle v\rangle \cdot \mathtt{out}_u \otimes \mathtt{def}_w(t)\, [\![P_1]\!] \not\searrow$, it follows (according to Lemma 2) that one of the following cases remains possible:

(1) $[\![P_2]\!] \searrow b'$ and $a' = \langle v\rangle \cdot \mathtt{out}_u \otimes b' \otimes \mathtt{def}_w(t)\, [\![P_1]\!]$;
(2) $[\![P_2]\!] \otimes \mathtt{def}_w(t)\, [\![P_1]\!] \searrow b'$ and $a' = \langle v\rangle \cdot \mathtt{out}_u \otimes b'$;
(3) $\langle v\rangle \cdot \mathtt{out}_u \otimes [\![P_2]\!] \searrow a''$ and $a' = a'' \otimes \mathtt{def}_w(t)\, [\![P_1]\!]$.

In case (1), $[\![P]\!] = \nu \cdot (w)([\![P_2]\!] \otimes \mathtt{def}_w(t)\, [\![P_1]\!]) \searrow \nu \cdot (w)(b' \otimes \mathtt{def}_w(t)\, [\![P_1]\!])$. Considering $b = \nu \cdot (w)(b' \otimes \mathtt{def}_w(t)\, [\![P_1]\!])$, it satisfies the requirements (according to Proposition 2). In case (2), we have $[\![P]\!] \searrow \nu \cdot (w)b'$, and consider $b = \nu \cdot (w)b'$. In case (3), by induction hypothesis, $[\![P_2]\!] \searrow b'$ and $a'' = \langle v\rangle \cdot \mathtt{out}_u \otimes b'$. Thus, $[\![P]\!] \searrow \nu \cdot (w)(b' \otimes \mathtt{def}_w(t)\, [\![P_1]\!])$, and consider $b = \nu \cdot (w)(b' \otimes \mathtt{def}_w(t)\, [\![P_1]\!])$. □

Lemma 7. $[\![P]\!] \otimes [\![Q]\!] \searrow a$ *iff one of the following conditions holds:*

1. $[\![P]\!] \searrow b$ *and* $a = b \otimes [\![Q]\!]$;
2. $[\![Q]\!] \searrow b$ *and* $a = [\![P]\!] \otimes b$.

Proof. ($\Leftarrow$) A consequence of the fact that the reaction is closed under tensorial product and equality.
($\Rightarrow$) Induction on the structure of P.

- If P is the empty process 0, then condition 2 holds obviously.
- If P is a message, then condition 2 holds by Lemma 6.
- If P is a parallel composition $P_1 \mid P_2$, then $[\![P_1]\!] \otimes [\![P_2]\!] \otimes [\![Q]\!] \searrow a$.

  By Lemma 2, it follows that one of the following cases is possible:

(i) $[\![Q]\!] \searrow b'$ and $a = [\![P_1]\!] \otimes [\![P_2]\!] \otimes b'$;
(ii) $[\![P_i]\!] \searrow b'$ and $a = b' \otimes [\![P_j]\!] \otimes [\![Q]\!]$;
(iii) $[\![P_i]\!] \otimes [\![P_j]\!] \searrow b'$ and $a = b' \otimes [\![Q]\!]$;
(iv) $[\![P_i]\!] \otimes [\![Q]\!] \searrow a'$ and $a = a' \otimes [\![P_j]\!]$, where $\{i,j\} = [2]$.

According to Proposition 2, $[\![P]\!] = [\![P_i]\!] \otimes [\![P_j]\!]$. In case ($i$), condition 2 holds by taking $b = b'$. In case (ii), we have $[\![P]\!] \searrow b' \otimes [\![P_j]\!]$. Then, condition *1* holds by taking $b = b' \otimes [\![P_j]\!]$. In case ($iii$), condition *1* holds by taking $b = b'$.
In case (iv), by induction, we distinguish two sub-cases:

(a) $[\![P_i]\!] \searrow b'$ and $a' = b' \otimes [\![Q]\!]$;
(b) $[\![Q]\!] \searrow b'$ and $a' = [\![P_i]\!] \otimes b'$.

For (a), we obtain $[\![P]\!] \searrow b' \otimes [\![P_j]\!]$, and condition *1* holds for $b = b' \otimes [\![P_j]\!]$. For ($b$), condition 2 holds for $b = b'$. In both sub-cases, some action commutations are required; they are possible according to Proposition 2.

- If P is a definition $\mathtt{def}\ w\langle t\rangle \triangleright P_1\ \mathtt{in}\ P_2$, then we may assume without losing generality that $w \notin \mathtt{fn}(Q)$. By Lemma 4, $w \notin \mathtt{surf}([\![Q]\!])$. According to Proposition 2, we have $\nu \cdot (w)([\![P_2]\!] \otimes \mathtt{def}_w(t)\, [\![P_1]\!] \otimes [\![Q]\!]) \searrow a$. By Lemma 2, $[\![P_2]\!] \otimes \mathtt{def}_w(t)\, [\![P_1]\!] \otimes [\![Q]\!] \searrow a'$ and $a = \nu \cdot (w)a'$. Since $\mathtt{def}_w(t)\, [\![P_1]\!] \not\searrow$, it follows from Lemma 2 that one of the following cases remains possible:

  (i) $[\![P_2]\!] \searrow b'$ and $a' = b' \otimes \mathtt{def}_w(t)\, [\![P_1]\!] \otimes [\![Q]\!]$;

(*ii*) $[\![Q]\!] \searrow b'$ and $a' = [\![P_2]\!] \otimes \mathtt{def}_w(t) [\![P_1]\!] \otimes b'$;
(*iii*) $[\![P_2]\!] \otimes \mathtt{def}_w(t) [\![P_1]\!] \searrow b'$ and $a' = b' \otimes [\![Q]\!]$;
(*iv*) $[\![P_2]\!] \otimes [\![Q]\!] \searrow a''$ and $a' = a'' \otimes \mathtt{def}_w(t) [\![P_1]\!]$.

In case (i), condition *1* holds for $b = \nu \cdot (w)(b' \otimes \mathtt{def}_w(t) [\![P_1]\!])$.

In case (ii), by Lemma 2, there exists b such that $\mathtt{surf}(b) \subseteq \mathtt{surf}([\![Q]\!])$ and $b = b'$; condition 2 holds for this b. In case (iii), condition *1* holds for $b = \nu \cdot (w)b'$.

In case (iv), we distinguish two sub-cases:

(*a*) $[\![P_2]\!] \searrow b'$ and $a'' = b' \otimes [\![Q]\!]$;
(*b*) $[\![Q]\!] \searrow b'$ and $a'' = [\![P_2]\!] \otimes b'$.

For (a), condition *1* holds for $b = \nu \cdot (w)(b' \otimes \mathtt{def}_w(t) [\![P_1]\!])$. For (b), there exists b such that $\mathtt{surf}(b) \subseteq \mathtt{surf}([\![Q]\!])$ and $b = b'$ (by Lemma 2); condition 2 holds for this b. Proposition 2 is used in all cases and sub-cases. □

Lemma 8. $[\![P]\!] \otimes \mathtt{def}_u(y) [\![Q]\!] \searrow a$ *iff one of the following conditions holds:*

1. $[\![P]\!] \searrow b$ *and* $a = b \otimes \mathtt{def}_u(y) [\![Q]\!]$;
2. $P \equiv R \mid u\langle v\rangle$ *and* $a = [\![R \mid \{v/y\}Q]\!] \otimes \mathtt{def}_u(y) [\![Q]\!]$;
3. $P \equiv \mathtt{def}\ v_1\langle t_1\rangle \triangleright R_1\ \mathtt{in}\ \mathtt{def}\ v_2\langle t_2\rangle \triangleright R_2\ \mathtt{in} \ldots \mathtt{def}\ v_n\langle t_n\rangle \triangleright R_n\ \mathtt{in}\ (R \mid u\langle v_n\rangle)$, *and*

$$\begin{aligned} a \;=\; & \nu \cdot (v_1)(\mathtt{def}_{v_1}(t_1) [\![R_1]\!] \otimes \\ & \quad \nu \cdot (v_2)(\mathtt{def}_{v_2}(t_2) [\![R_2]\!] \otimes \\ & \quad\quad \vdots \\ & \quad\quad \nu \cdot (v_n)(\mathtt{def}_{v_n}(t_n) [\![R_n]\!] \otimes [\![R \mid \{v_n/y\}Q]\!] \otimes \mathtt{def}_u(y) [\![Q]\!]) \ldots)), \end{aligned}$$

where $v_i \notin \mathtt{fn}(Q) \cup \{u\}$ *for every* $i \in [n]$.

Proof. $(\Leftarrow)$ If condition *1* holds, then the implication follows as a consequence of the fact that the reaction is closed under tensorial product and equality. If condition *2* holds, then we have

$$\begin{aligned} & [\![P]\!] \otimes \mathtt{def}_u(y) [\![Q]\!] = && \text{by Proposition 5} \\ = \;& [\![R]\!] \otimes \langle v\rangle \cdot \mathtt{out}_u \otimes \mathtt{def}_u(y) [\![Q]\!] && \text{by Lemma 2} \\ \searrow \;& [\![R]\!] \otimes [v/y] [\![Q]\!] \otimes \mathtt{def}_u(y) [\![Q]\!] && \text{by Lemma 5} \\ = \;& a. \end{aligned}$$

If condition *3* holds, it follows by Lemma 2 and Lemma 4 that $v_i \notin \mathtt{surf}(\mathtt{def}_u(y) [\![Q]\!])$ for every $i \in [n]$. Then,

$$\begin{aligned} & [\![P]\!] \otimes \mathtt{def}_u(y) [\![Q]\!] = && \text{by Propositions 5 and 2} \\ = \;& \nu \cdot (v_1)(\mathtt{def}_{v_1}(t_1) [\![R_1]\!] \otimes && \text{by Lemma 2} \\ & \quad \vdots \\ & \nu \cdot (v_n)(\mathtt{def}_{v_n}(t_n) [\![R_n]\!] \otimes [\![R]\!] \otimes \langle v_n\rangle \cdot \mathtt{out}_u \otimes \mathtt{def}_u(y) [\![Q]\!]) \ldots) \\ & \searrow \;\nu \cdot (v_1)(\mathtt{def}_{v_1}(t_1) [\![R_1]\!] \otimes && \text{by Lemma 5} \\ & \quad \vdots \\ & \quad \nu \cdot (v_n)(\mathtt{def}_{v_n}(t_n) [\![R_n]\!] \otimes [\![R]\!] \otimes [v_n/y] [\![Q]\!] \otimes \mathtt{def}_u(y) [\![Q]\!]) \ldots) \\ & = \; a. \end{aligned}$$

$(\Rightarrow)$ Induction on the structure of P.

– If P is the empty process 0 or a message $w\langle v\rangle$ with $w \neq u$, then $[\![P]\!] \otimes \mathtt{def}_u(y) [\![Q]\!] \not\searrow$. The statement of the lemma is obviously true as its premise is not satisfied. On the other hand, if P is a message $u\langle v\rangle$, then

$$\begin{aligned} & [\![P]\!] \otimes \mathtt{def}_u(y) [\![Q]\!] \searrow && \text{by Lemma 2} \\ \searrow \;& [v/y] [\![Q]\!] \otimes \mathtt{def}_u(y) [\![Q]\!] \searrow a && \text{by Lemma 5} \\ = \;& 0 \mid \{v/y\}Q \otimes \mathtt{def}_u(y) [\![Q]\!]. \end{aligned}$$

Furthermore, $P \equiv 0 \mid u\langle v\rangle$. Consequently, condition *2* holds.

– If P is a parallel composition $P_1 \mid P_2$, then $[\![P_1]\!] \otimes [\![P_2]\!] \otimes \mathtt{def}_u(y) [\![Q]\!] \searrow a$. Since $\mathtt{def}_u(y) [\![Q]\!] \not\searrow$, it follows from Lemma 2 that one of the following cases remains possible:

(i) $[\![P_i]\!] \searrow b'$ and $a = b' \otimes [\![P_j]\!] \otimes \mathtt{def}_u\,(y)\,[\![Q]\!]$,
(ii) $[\![P_i]\!] \otimes [\![P_j]\!] \searrow b'$ and $a = b' \otimes \mathtt{def}_u\,(y)\,[\![Q]\!]$,
(iii) $[\![P_i]\!] \otimes \mathtt{def}_u\,(y)\,[\![Q]\!] \searrow a'$ and $a = a' \otimes [\![P_j]\!]$, where $\{i, j\} = [2]$.

According to Proposition 2, we obtain $[\![P]\!] = [\![P_i]\!] \otimes [\![P_j]\!]$. In case (i), we obtain $[\![P]\!] \searrow b' \otimes [\![P_j]\!]$, and condition *1* holds for $b = b' \otimes [\![P_j]\!]$. In case (ii), condition *1* holds for $b = b'$. In case (iii), we distinguish three sub-cases:

(a) $[\![P_i]\!] \searrow b'$ and $a' = b' \otimes \mathtt{def}_u\,(y)\,[\![Q]\!]$;
(b) $P_i \equiv R' \mid u\langle v\rangle$ and $a' = [\![R' \mid \{v/y\}Q]\!] \otimes \mathtt{def}_u\,(y)\,[\![Q]\!]$;
(c) $P_i \equiv \mathtt{def}\ v_1\langle t_1\rangle \rhd R_1\ \mathtt{in}\ \ldots\ \mathtt{def}\ v_n\langle t_n\rangle \rhd R_n\ \mathtt{in}\ (R' \mid u\langle v_n\rangle)$ and

$$\begin{aligned} a' &= \nu\cdot(v_1)(\mathtt{def}_{v_1}\,(t_1)\,[\![R_1]\!] \otimes \\ &\qquad \vdots \\ &\qquad \nu\cdot(v_n)(\mathtt{def}_{v_n}\,(t_n)\,[\![R_n]\!] \otimes [\![R' \mid \{v_n/y\}Q]\!] \otimes \mathtt{def}_u\,(y)\,[\![Q]\!])\ldots), \end{aligned}$$

where $v_k \notin \mathtt{fn}(Q) \cup \{u\}$ for every $k \in [n]$.

In sub-case (a), we obtain $[\![P]\!] \searrow b' \otimes [\![P_j]\!]$. Therefore, condition *1* holds for $b = b' \otimes [\![P_j]\!]$. For (b), we have $P \equiv R' \mid P_j \mid u\langle v\rangle$. By Proposition 2, $a = [\![R' \mid P_j \mid \{v/y\}Q]\!] \otimes \mathtt{def}_u\,(y)\,[\![Q]\!]$. Thus, condition 2 holds. For sub-case (c), we may assume (without losing generality) that $v_k \notin \mathtt{fn}(P_j)$ for every $k \in [n]$. By Lemma 4, it follows that $v_k \notin \mathtt{surf}(\,[\![P_j]\!]\,)$ for every $k \in [n]$. Then $P \equiv \mathtt{def}\ v_1\langle t_1\rangle \rhd R_1\ \mathtt{in}\ \ldots\ \mathtt{def}\ v_n\langle t_n\rangle \rhd R_n\ \mathtt{in}\ (R' \mid P_j \mid u\langle v_n\rangle)$, and

$$\begin{aligned} a &= \qquad\qquad\qquad\qquad \text{by Proposition 2} \\ &= \nu\cdot(v_1)(\mathtt{def}_{v_1}\,(t_1)\,[\![R_1]\!] \otimes \\ &\qquad \vdots \\ &\qquad \nu\cdot(v_n)(\mathtt{def}_{v_n}\,(t_n)\,[\![R_n]\!] \otimes [\![R' \mid P_j \mid \{v_n/y\}Q]\!] \otimes \mathtt{def}_u\,(y)\,[\![Q]\!])\ldots). \end{aligned}$$

Thus, condition *3* holds.

– If P is a definition $\mathtt{def}\ w\langle t\rangle \rhd P_1\ \mathtt{in}\ P_2$, then we can assume without losing generality that $w \notin \mathtt{fn}(Q) \cup \{u\}$. By Lemma 4 and Lemma 2, $w \notin \mathtt{surf}(\mathtt{def}_u\,(y)\,[\![Q]\!]\,)$. It follows from Proposition 2 that $[\![P]\!] \otimes \mathtt{def}_u\,(y)\,[\![Q]\!] = \nu\cdot(w)(\,[\![P_2]\!] \otimes \mathtt{def}_w\,(t)\,[\![P_1]\!] \otimes \mathtt{def}_u\,(y)\,[\![Q]\!]\,) \searrow a$. By Lemma 2, $[\![P_2]\!] \otimes \mathtt{def}_w\,(t)\,[\![P_1]\!] \otimes \mathtt{def}_u\,(y)\,[\![Q]\!] \searrow a'$ and $a = \nu\cdot(w)a'$. Since $\mathtt{def}_w\,(t)\,[\![P_1]\!] \not\searrow$ and $\mathtt{def}_u\,(y)\,[\![Q]\!] \not\searrow$, then $\mathtt{def}_u\,(y)\,[\![Q]\!] \otimes \mathtt{def}_w\,(t)\,[\![P_1]\!] \not\searrow$ (according to Lemma 2). It follows that one of the following cases remains possible:

(i) $[\![P_2]\!] \searrow b'$ and $a' = b' \otimes \mathtt{def}_w\,(t)\,[\![P_1]\!] \otimes \mathtt{def}_u\,(y)\,[\![Q]\!]$;
(ii) $[\![P_2]\!] \otimes \mathtt{def}_w\,(t)\,[\![P_1]\!] \searrow b'$ and $a' = b' \otimes \mathtt{def}_u\,(y)\,[\![Q]\!]$;
(iii) $[\![P_2]\!] \otimes \mathtt{def}_u\,(y)\,[\![Q]\!] \searrow a''$ and $a' = a'' \otimes \mathtt{def}_w\,(t)\,[\![P_1]\!]$.

In case (i), condition *1* holds for $b = \nu\cdot(w)(b' \otimes \mathtt{def}_w\,(t)\,[\![P_1]\!]\,)$. In case (ii), condition *1* holds for $b = \nu\cdot(w)b'$. In case (iii), by induction, we distinguish three sub-cases:

(a) $[\![P_2]\!] \searrow b'$ and $a'' = b' \otimes \mathtt{def}_u\,(y)\,[\![Q]\!]$,
(b) $P_2 \equiv R' \mid u\langle v\rangle$ and $a'' = [\![R' \mid \{v/y\}Q]\!] \otimes \mathtt{def}_u\,(y)\,[\![Q]\!]$,
(c) $P_2 \equiv \mathtt{def}\ v_1\langle t_1\rangle \rhd R_1\ \mathtt{in}\ \ldots\ \mathtt{def}\ v_n\langle t_n\rangle \rhd R_n\ \mathtt{in}\ (R' \mid u\langle v_n\rangle)$ and

$$\begin{aligned} a'' &= \nu\cdot(v_1)(\mathtt{def}_{v_1}\,(t_1)\,[\![R_1]\!] \otimes \\ &\qquad \vdots \\ &\qquad \nu\cdot(v_n)(\mathtt{def}_{v_n}\,(t_n)\,[\![R_n]\!] \otimes [\![R' \mid \{v_n/y\}Q]\!] \otimes \mathtt{def}_u\,(y)\,[\![Q]\!])\ldots), \end{aligned}$$

where $v_k \notin \mathtt{fn}(Q) \cup \{u\}$ for every $k \in [n]$.

In sub-case (a), condition *1* holds for $b = \nu\cdot(w)(b' \otimes \mathtt{def}_w\,(t)\,[\![P_1]\!]\,)$. In sub-case (b), we have $P \equiv \mathtt{def}\ w\langle t\rangle \rhd P_1\ \mathtt{in}\ (R' \mid u\langle v\rangle)$. We distinguish two situations:

- $v \neq w$. Then $P \equiv \mathtt{def}\ w\langle t\rangle \rhd P_1\ \mathtt{in}\ R' \mid u\langle v\rangle$. It is easy to see that $\mathtt{surf}(\,[\![\{v/y\}Q]\!]\,) \subseteq \{v\} \cup \mathtt{surf}(\,[\![Q]\!]\,)$, and so $w \notin \mathtt{surf}(\,[\![\{v/y\}Q]\!]\,)$. By Proposition 2, $a = [\![\mathtt{def}\ w\langle t\rangle \rhd P_1\ \mathtt{in}\ R' \mid \{v/y\}Q]\!] \otimes \mathtt{def}_u\,(y)\,[\![Q]\!]$. Thus, condition 2 holds.
- $v = w$. Then $P \equiv \mathtt{def}\ v\langle t\rangle \rhd P_1\ \mathtt{in}\ (R' \mid u\langle v\rangle)$. Moreover, $a = \nu\cdot(v)(\mathtt{def}_v\,(t)\,[\![P_1]\!] \otimes [\![R' \mid \{v/y\}Q]\!] \otimes \mathtt{def}_u\,(y)\,[\![Q]\!]\,)$. Thus, condition 3 holds.

In sub-case (c), $P \equiv \texttt{def}\ w\langle t\rangle \triangleright P_1\ \texttt{in}\ \texttt{def}\ v_1\langle t_1\rangle \triangleright R_1\ \texttt{in}\ \ldots\ \texttt{def}\ v_n\langle t_n\rangle \triangleright R_n\ \texttt{in}\ (R' \mid u\langle v_n\rangle)$. Using Proposition 2, we obtain

$$\begin{aligned} a &= \nu\cdot(w)(\texttt{def}_w\,(t)\,[\![\,P_1\,]\!] \otimes \\ &\quad \nu\cdot(v_1)(\texttt{def}_{v_1}\,(t_1)\,[\![\,R_1\,]\!] \otimes \\ &\quad\quad \vdots \\ &\quad \nu\cdot(v_n)(\texttt{def}_{v_n}\,(t_n)\,[\![\,R_n\,]\!] \otimes [\![\,R' \mid \{v_n/y\}Q\,]\!] \otimes \texttt{def}_u\,(y)\,[\![\,Q\,]\!]\,)\ldots)). \end{aligned}$$

Thus, condition 3 holds. □

Theorem 2. *If* $[\![\,P\,]\!] \searrow a$*, then there exists a process* Q *such that* $P \to Q$ *and* $[\![\,Q\,]\!] = a$.

Proof. Induction on the structure of P.

– If P is the empty process or a message, then $[\![\,P\,]\!] \not\searrow$. Therefore, the statement of the theorem is obviously true because the premise is not satisfied.

– If P is a parallel composition $P_1 \mid P_2$, then $[\![\,P_1\,]\!] \otimes [\![\,P_2\,]\!] \searrow a$. By Lemma 7, one of the following cases holds:

(1) $[\![\,P_1\,]\!] \searrow a_1$ and $a = a_1 \otimes [\![\,P_2\,]\!]$;
(2) $[\![\,P_2\,]\!] \searrow a_2$ and $a = [\![\,P_1\,]\!] \otimes a_2$.

It is sufficient to consider the case (1), the other one being similar (symmetric).

By induction, we have $P_1 \to Q_1$ and $a_1 = [\![\,Q_1\,]\!]$. According to Proposition 1, $P \to Q_1 \mid P_2$ and $a = [\![\,Q_1\,]\!] \otimes [\![\,P_2\,]\!]$. Thus, the result of the theorem holds for $Q = Q_1 \mid P_2$.

– If P is a definition $\texttt{def}\ w\langle t\rangle \triangleright P_1\ \texttt{in}\ P_2$, then $\nu\cdot(w)(\,[\![\,P_2\,]\!] \otimes \texttt{def}_w\,(t)\,[\![\,P_1\,]\!] \searrow a$. It follows from Lemma 2 that $[\![\,P_2\,]\!] \otimes \texttt{def}_w\,(t)\,[\![\,P_1\,]\!] \searrow a'$ and $a = \nu\cdot(w)a'$. By Lemma 8, only one of the following cases holds:

(i) $[\![\,P_2\,]\!] \searrow b$ and $a' = b \otimes \texttt{def}_w\,(t)\,[\![\,P_1\,]\!]$;
(ii) $P_2 \equiv R \mid w\langle v\rangle$ and $a' = [\![\,R \mid \{v/t\}P_1\,]\!] \otimes \texttt{def}_w\,(t)\,[\![\,P_1\,]\!]$;
(iii) $P_2 \equiv \texttt{def}\ w_1\langle t_1\rangle \triangleright R_1\ \texttt{in}\ \ldots\ \texttt{def}\ w_n\langle t_n\rangle \triangleright R_n\ \texttt{in}\ (R \mid w\langle w_n\rangle)$ and

$$\begin{aligned} a' &= \nu\cdot(w_1)(\texttt{def}_{w_1}\,(t_1)\,[\![\,R_1\,]\!] \otimes \\ &\quad\quad \vdots \\ &\quad \nu\cdot(w_n)(\texttt{def}_{w_n}\,(t_n)\,[\![\,R_n\,]\!] \otimes [\![\,R \mid \{w_n/t\}P_1\,]\!] \otimes \texttt{def}_w\,(t)\,[\![\,P_1\,]\!]\,)\ldots), \end{aligned}$$

where $w_i \notin \texttt{fn}(P_1) \cup \{w\}$ for every $i \in [n]$.

In case (i), by the induction hypothesis, $P_2 \to Q_2$ and $b = [\![\,Q_2\,]\!]$. It follows that $P \to \texttt{def}\ w\langle t\rangle \triangleright P_1\ \texttt{in}\ Q_2$ and $a = \nu\cdot(w)(\,[\![\,Q_2\,]\!] \otimes \texttt{def}_w\,(t)\,[\![\,P_1\,]\!]\,)$ Thus, the result of the theorem holds for $Q = \texttt{def}\ w\langle t\rangle \triangleright P_1\ \texttt{in}\ Q_2$.

In case (ii), we have $P \to \texttt{def}\ w\langle t\rangle \triangleright P_1\ \texttt{in}\ (R \mid \{v/t\}P_1)$ and $a = \nu\cdot(w)(\,[\![\,R \mid \{v/t\}P_1\,]\!] \otimes \texttt{def}_w\,(t)\,[\![\,P_1\,]\!]\,) = [\![\,Q\,]\!]$. Thus, the result of the theorem holds for $Q = \texttt{def}\ w\langle t\rangle \triangleright P_1\ \texttt{in}\ (R \mid \{v/t\}P_1)$.

In case (iii), it follows that $w_i \notin \texttt{surf}(\texttt{def}_w\,(t)\,[\![\,P_1\,]\!]\,)$ for any $i \in [n]$ (Lemmas 2 and 4).

$$P \equiv \texttt{def}\ w\langle t\rangle \triangleright P_1\ \texttt{in}\ \texttt{def}\ w_1\langle t_1\rangle \triangleright R_1\ \texttt{in}\ \ldots\texttt{def}\ w_n\langle t_n\rangle \triangleright R_n\ \texttt{in}\ (R \mid w\langle w_n\rangle) \to \underbrace{\texttt{def}\ w\langle t\rangle \triangleright P_1\ \texttt{in}\ \texttt{def}\ w_1\langle t_1\rangle \triangleright R_1\ \texttt{in}\ \ldots\ \texttt{def}\ w_n\langle t_n\rangle \triangleright R_n\ \texttt{in}\ (R \mid \{w_n/t\}P_1)}_{Q}$$

$$\begin{aligned} a &= \nu\cdot(w)(&& \text{by Proposition 2} \\ &\quad \nu\cdot(w_1)(\texttt{def}_{w_1}\,(t_1)\,[\![\,R_1\,]\!] \otimes \\ &\quad\quad \vdots \\ &\quad \nu\cdot(w_n)(\texttt{def}_{w_n}\,(t_n)\,[\![\,R_n\,]\!] \otimes [\![\,R \mid \{w_n/t\}P_1\,]\!] \otimes \texttt{def}_w\,(t)\,[\![\,P_1\,]\!]\,)\ldots)) \\ &= \nu\cdot(w)(\texttt{def}_w\,(t)\,[\![\,P_1\,]\!] \otimes \\ &\quad \nu\cdot(w_1)(\texttt{def}_{w_1}\,(t_1)\,[\![\,R_1\,]\!] \otimes \\ &\quad\quad \vdots \\ &\quad \nu\cdot(w_n)(\texttt{def}_{w_n}\,(t_n)\,[\![\,R_n\,]\!] \otimes [\![\,R \mid \{w_n/t\}P_1\,]\!]\,)\ldots)) \\ &= [\![\,Q\,]\!]\,. \end{aligned}$$

□

6. Describing Communication Patterns by Using the Hypergraph Model

In the Unix operating system, interprocess communications based on message queues allow exchange of information between processes. The processes exchange information by accessing a common message queue. Essentially, one process produces a message queue (via a message-passing module) that other processes may access; often a server places a message onto a queue which can be read by multiple clients. The sending process may specify its type when placing the message in a queue such that the reading processes can select the appropriate message; thus, message queues provide a way of multiplexing information from one producer to more consumers.

As example, we consider a simple system in which only one channel is used to exchange messages between the server and clients, and any message at the input of any client must appear at the output of all the clients (this is a requirement for several social networks including a chat messaging system). A type associated to each message allows a client to access the (unique) message queue for selectively reading only specific messages (in a first-in–first-out manner). We simplify the system, and consider a process S working as a server and two clients A and B. The channels `idA` and `idB` are used to indicate the type of messages from S to A and B, respectively; a channel `idS` indicates the type of messages from the clients to the server. Client A uses an input channel `inA` and an output channel **outA**; client B uses channels `inB` and **outB**). For a message `m` sent along the input channel `inA`, the pattern calculus process corresponding to this system is:

$$\begin{array}{llrcll} CommSyst = & \texttt{def} & q\langle x\rangle \mid idS\langle y\rangle & \triangleright & q\langle x\rangle \mid idA\langle\_\rangle \mid q\langle x\rangle \mid idB\langle\_\rangle & \wedge \\ & & q\langle x\rangle \mid idA\langle y\rangle & \triangleright & outA\langle x\rangle & \wedge \\ & & q\langle x\rangle \mid idB\langle y\rangle & \triangleright & outB\langle x\rangle & \wedge \\ & & inA\langle x\rangle & \triangleright & q\langle x\rangle \mid idS\langle\_\rangle & \wedge \\ & & inB\langle x\rangle & \triangleright & q\langle x\rangle \mid idS\langle\_\rangle & \\ & \texttt{in} & inA\langle m\rangle \,. & & & \end{array}$$

Using the hypergraph model, in Figure 3 is presented the net corresponding to this process.

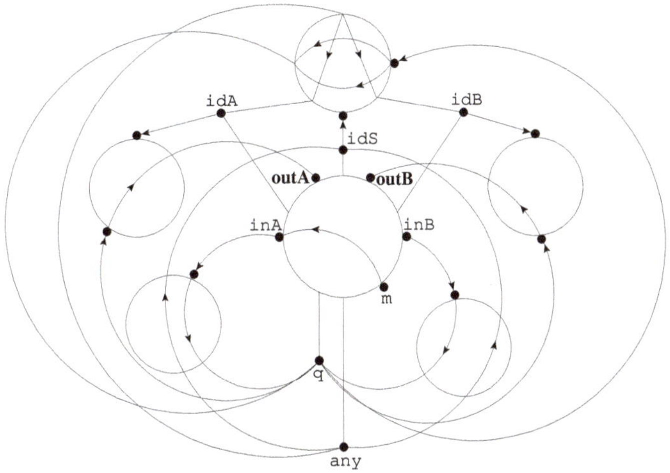

Figure 3. The net of a simple communication system described previously in pattern calculus.

Except the root hyperedge, the structure of this net does not change during the evolution. Therefore, the evolution of the system could be described graphically focusing only on the root hyperedge; this evolution is depicted in Figure 4.

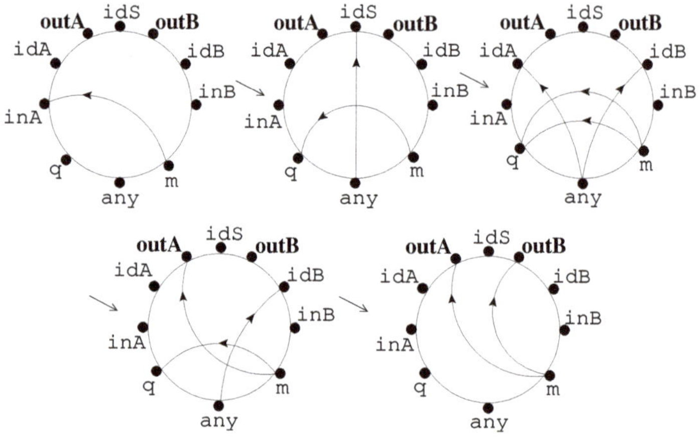

Figure 4. The evolution of the system (as it appears in the root hyperedge).

In Figure 4 it is not difficult to check visually the requirement that a message appearing at the input of a client appears also at the output of all the clients. In our case, the message `m` on the input channel `inA` (the initial step) appears at the output channels **outA** and **outB** in the final step described in Figure 4.

7. Conclusions and Related Work

In this paper we introduce a hypergraph model (given by the pattern nets) for the communication patterns. These nets provide a fully abstract model for the pattern calculus. In this way, a new sound graphical model for concurrency is introduced. We present a semantic interpretation of the pattern calculus in the framework of control structures, creating a graphical representation for the pattern calculus given by a new hypergraph model given by the pattern nets. By introducing a mapping from the control structure of pattern calculus into a set of hypergraphs, we provide a graphical model for communication patterns. It is also proved that the hypergraph model preserves the operational reductions of processes from pattern calculus and of the actions from the control structures. As an example, simple interprocess communications based on message queues inspired by the social networks are described by using our pattern nets. This example could be a first step towards more realistic scenarios in which the proposed model can be used to identify control structures supporting specific communication patterns. Future work will investigate realistic autonomic networking, mobility management, multiaccess selection, wireless and mobile networks (as they are presented in [12], for instance).

Graphical representations for process calculi highlight a new perception, providing a visual approach of concurrency and networks. According to our knowledge, just a few papers are devoted to the graphical presentations of the process calculi. We mention our previous attempts, namely the faithful π-nets [13], a graphical representation of the π-calculus machine [14], and a related approach by using jc-nets [15]. There exist also the graphical representations introduced by Robin Milner, namely action graphs and π-nets. Action graphs [16] are the graphical presentation of action calculi; they are very general, and so they are not able to describe specific features of certain action calculi. In the graphical presentation of the π-calculus given by the π-nets [6], channels are represented as rather complicated nodes called torpedos together with boxes representing guards, and messages are represented as directed arcs. The boxes obscure the internal nodes representing channels; to ensure access to the hidden channels, a rather complex additional mechanism of links is used. To avoid such a mechanism, in [13,14] the channels are represented by nodes, messages are represented by boxes of arcs and guards are represented by arcs between boxes. This approach simplified the graphical representation of the π-calculus; unfortunately, it provided identical representations for processes with different behaviours. Fortunately, this deficiency was overtaken in the pattern calculus hypergraph model: processes with different behaviours are not mapped to the same hypergraph. The hypergraph model is presented in the same formal framework used for the π-nets (it is worth noting that hypergraph model avoids certain irrelevant aspects of π-nets). It is simpler than the π-nets, preserving much of their expressive power (according to [2], the join calculus has the same expressive power as the π-calculus). Compared with all of them, the pattern nets represent a simple but sound graphical model for concurrency, providing a fully abstract model for the pattern calculus.

Funding: This research received no external funding.

Acknowledgments: Many thanks to Mihai Rotaru for his contributions in our past collaboration.

Conflicts of Interest: The author declares no conflict of interest.

References

1. Resnik, M.D. *Mathematics as a Science of Patterns*; Oxford University Press: Oxford, UK, 1999.
2. Fournet, C; Gonthier, G. The reflexive CHAM and the join calculus. In Proceedings of the 23rd ACM Symposium on Principles of Programming Languages (POPL'96), St. Petersburg, FL, USA; 21–24 January 1996; Association for Computing Machinery: New York, NY, USA, 1996; pp. 372–385. [CrossRef]
3. Levy, J.J. Some results in the join calculus. *Lect. Notes Comput. Sci.* **1997**, *1281*, 233–249.
4. Milner, R. *Communicating and Mobile Systems: The π-Calculus*; Cambridge University Press: Cambridge, UK, 1999.
5. Reisig, W. *Understanding Petri Nets*; Springer: Berlin, Germany, 2013.
6. Milner, R. π-nets: A graphical form of π-calculus. *Lect. Notes Comput. Sci.* **1994**, *788*, 26–42.
7. Schmidt, G.; Strohlein, T. *Relations and Graphs*; EATCS Monographs on Theor. Comput. Sci. Springer: Berlin, Germany, 1993.
8. Milner, R. Action calculi for syntactic action structures. *Lect. Notes Comput. Sci.* **1993**, *711*, 105–121.
9. Mifsud, A.; Milner, R.; Power, J. Control structures. In Proceedings of the 10th IEEE Symposium on Logic in Computer Science (LICS'95), San Diego, CA, USA, 26–29 June 1995.
10. Asperti, A.; Longo, G. *Categories, Types and Structures*; MIT Press: Cambridge, MA, USA, 1996.
11. Milner, R. Fully abstract models of typed λ-calculi. *Theor. Comput. Sci.* **1977**, *4*, 1–22. [CrossRef]
12. Pentikousis, K.; Blume, O.; Aguero, R.; Papavassiliou, S. *Mobile Networks and Management*; Springer: Berlin, Germany, 2010.
13. Ciobanu G.; Rotaru, M. Faithful π-nets. A graphical representation of the asynchronous π-calculus. *Electron. Notes Theor. Comput. Sci.* **1998**, *18*, 24–45. [CrossRef]
14. Ciobanu G.; Rotaru, M. A π-calculus machine. *J. Univers. Comput. Sci.* **2000**, *6*, 39–59.
15. Ciobanu G.; Rotaru, M. JC-nets. *Lect. Notes Comput. Sci.* **2001**, *2055*, 190–201.
16. Milner, R. Calculi for interaction. *Acta Inform.* **1996**, *33*, 707–737. [CrossRef]

axioms

MDPI

Article

Logarithmic SAT Solution with Membrane Computing

Radu Nicolescu *,†, Michael J. Dinneen, James Cooper, Alec Henderson and Yezhou Liu

School of Computer Science, University of Auckland, Private Bag 92019, Auckland 1142, New Zealand; mjd@cs.auckland.ac.nz (M.J.D.); jcoo092@aucklanduni.ac.nz (J.C.); ahen386@aucklanduni.ac.nz (A.H.); yliu442@aucklanduni.ac.nz (Y.L.)

* Correspondence: r.nicolescu@auckland.ac.nz

† As a humble commemoration of Professor Solomon Marcus' fifth death anniversary.

Abstract: P systems have been known to provide efficient polynomial (often linear) deterministic solutions to hard problems. In particular, cP systems have been shown to provide very crisp and efficient solutions to such problems, which are typically linear with small coefficients. Building on a recent result by Henderson et al., which solves SAT in square-root-sublinear time, this paper proposes an orders-of-magnitude-faster solution, running in logarithmic time, and using a small fixed-sized alphabet and ruleset (25 rules). To the best of our knowledge, this is the fastest deterministic solution across all extant P system variants. Like all other cP solutions, it is a complete solution that is not a member of a uniform family (and thus does not require any preprocessing). Consequently, according to another reduction result by Henderson et al., cP systems can also solve *k*-colouring and several other NP-complete problems in logarithmic time.

Keywords: membrane computing; P systems; cP systems; NP-complete; NP-hard; SAT; logarithmic time complexity

MSC: 68Q07; 68Q10; 68N17; 68W10

Citation: Nicolescu, R.; Dinneen, M.J.; Cooper, J.; Henderson, A.; Liu, Y. Logarithmic SAT Solution with Membrane Computing. *Axioms* **2022**, *11*, 66. https://doi.org/10.3390/axioms11020066

Academic Editor: Gexiang Zhang

Received: 15 January 2022
Accepted: 29 January 2022
Published: 8 February 2022

1. Introduction

The P-versus-NP problem remains one of the most important unsolved problems in computational complexity theory. Loosely following Sipser [1] and keeping the discussion focused on *deterministic algorithms*—as we do throughout this paper—the class P can be viewed as the class of decision problems that can be *solved* "quickly", whereas NP can be viewed as the possibly larger class of decision problems with solutions that can be *verified* "quickly", where "quickly" is taken in the theoretical sense, i.e., polynomial time. It is straightforward to see that $P \subseteq NP$. However, it is still unknown whether this inclusion is strict or not, in other words, whether $P \subsetneq NP$ or $P = NP$. In a nutshell, the big theoretical question is whether every problem of which the solution can be verified in polynomial time (NP) can also be solved in polynomial time (P).

The current widespread opinion is that $P \subsetneq NP$, as there are quite a few "hard" problems that can be "quickly" verified, but do not seem to have "quick" solutions, with their fastest known solutions taking time substantially greater than any polynomial (e.g., exponential). Therefore, many studies have investigated different approaches to solve such hard problems in a reasonable amount of time (e.g., polynomial or even linear time). Such methods include approximation [2], fixing parameters [3], or the use of alternative theoretical models, such as P systems [4–9].

P systems—also known as membrane computing—are a family of parallel and distributed biologically inspired models of computing, proposed by Gheorghe Păun in [10], first as cell-like P systems, then followed by many variants, such as P systems with active membranes [11], tissue-like P systems [12], neural-like P systems [13], and P systems with compound terms (cP systems) [14,15]. These systems have been found to have theoretically time-efficient solutions to many hard problems, even beyond NP, e.g., in PSPACE [16–21].

It may be worthwhile to note that, with the exception of cP systems, most other P systems solutions are actually *uniform families* of related solutions, with one custom solution (e.g., custom alphabet and ruleset) for each problem size n. Here, uniform means that each custom size n solution is built via an additional preprocessing phase, by means of an ad-hoc polynomial-time algorithm (typically not described but reasonably evident). In contrast, cP solutions are given by *fixed-size* alphabets and rulesets (typically small), while running with the same theoretical efficiency, or even faster.

In this work, we present a novel *deterministic* cP solution to SAT, running in *logarithmic time*, $\mathcal{O}(\log n)$. To the best of our knowledge, this represents a significant breakthrough in membrane computing, being orders-of-magnitude faster than all previous *deterministic* solutions. As mentioned, we do not consider here the interesting area of non-deterministic computations, where there are several interesting results, e.g., using neural-like P systems [22].

Our novel solution builds upon and substantially improves the already very fast cP solution to SAT recently proposed by Henderson et al. [4], which runs in square-root time, $\mathcal{O}(\sqrt{n})$. The solution presented here is based on a fast method of creating and evaluating a complete binary tree of height n, in $\mathcal{O}(\log n)$ time. When measuring the number of rule templates, we see that our new solution is comparable to those of previous P systems studies. However, when counting rules rather than the templates, we see that other solutions can have an exponential number of rules.

Using the results presented in this paper, reductions such as those presented in Stamm-Wilbrandt [23] and Henderson et al. [4] will enable more logarithmic time solutions, $\mathcal{O}(\log n)$, to quite a few other NP-complete problems, such as k-colouring.

However, to the best of our knowledge, all these efficient solutions are still theoretical and have not yet been practically implemented. Designing efficient, practical implementations is a topic of current research.

2. Background

In this section, we briefly recall the well-known Boolean satisfiability problem (SAT) and we offer a short introduction to cP systems.

2.1. The SAT Problem

SAT is one of the best-known examples of an NP-complete problem and is a relatively simple but central problem in many areas of computer science (e.g., complexity, artificial intelligence, cryptography, etc.). Like all other NP-complete problems, it has no known (worst-case) polynomial solution in the Turing machine model (or related models). In this paper, we show that cP systems can theoretically solve SAT in sublinear logarithmic time.

SAT determines if the variables of a given Boolean formula can be assigned Boolean values that evaluate the formula to true. A Boolean formula is an expression involving Boolean variables and Boolean operations. A Boolean formula is in conjunctive normal form (CNF) if it is expressed as a conjunction ($\wedge$) of clauses. A clause is a disjunction ($\vee$) of literals. A literal is a variable or its negation (here indicated by overbars).

Basic SAT assumes that the formulae are given in CNF, with implicit existential quantifiers on all variables. The existential quantifier ($\exists$) results are true if one of the possible assignments of the variables allows the formula to be true.

Example 1. *For example, the following Boolean formula with two variables is in CNF:*

$$(x_1 \vee x_2) \wedge (\overline{x_1} \vee \overline{x_2}).$$

SAT interprets the above formula as the following decision problem:

$$\exists x_1 \exists x_2 (x_1 \vee x_2) \wedge (\overline{x_1} \vee \overline{x_2}).$$

The size of the problem is given by the number of variables, n, e.g., $n = 2$, for the formula of Example 1. There are straightforward bijections between several sets of size 2^n: (i) candidate solutions (variable allocations) of a CNF formula with n variables, (ii) (characteristic functions for) subsets of set $\{1, 2, \ldots, n\}$, (iii) branches (root-to-leaf paths) of the complete binary tree of height n. For case (iii), a tree path starting from a root can be naturally labelled as a string of bits, where bits indicates its left/right "choices" (turns) in the top-down (root-to-leaf) order.

Example 2. *Consider the complete binary tree of Figure 1 of height $n = 2$, with 4 branches, in left-to-right order:*

Branch	*SectionAllocations*	*Subset*
00	$x_1 = 0, x_2 = 0$	*{} (empty)*
01	$x_1 = 0, x_2 = 1$	*{ 2 }*
10	$x_1 = 1, x_2 = 0$	*{ 1 }*
11	$x_1 = 1, x_2 = 1$	*{ 1, 2 }*

Note that branches 01 and 10 correspond to solutions for the formula of Example 1.

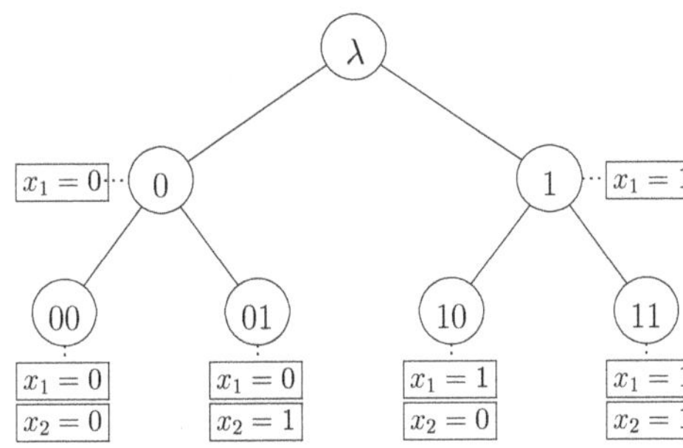

Figure 1. Complete binary tree of height 2. Nodes hold branch labels, and are decorated with attributes that are explicit corresponding variable allocations. Branches 01 and 10 correspond to solutions for the formula of Example 1, $(x_1 \vee x_2) \wedge (\overline{x_1} \vee \overline{x_2})$.

Our cP solution is based on a parallel construction of complete binary tree branches, followed by a parallel formula evaluation on these branches.

2.2. cP Systems

In this paper we propose a novel cP solution to a hard problem; to the best of our knowledge, this is the first P solution running in logarithmic time, which represents an improvement of orders of magnitude.

P systems, also known as membrane computing, are a framework for designing computational models inspired by biology. Similarly to many other P systems variants, such as cell-like and tissue-like P systems, cP systems are based on *nested labelled multisets* and offer: (i) unbounded access to resources, such as space and processing power; (ii) top-level cells, with sub-cells organised into nested tree structures; (iii) graph based networks of top-level cells; and (iv) evolutions driven by formal multiset rewriting rules, with additional messaging primitives between top-level cells.

However, distinctively, cP systems' multiset rewriting rules are *generic*, with *variables instantiated* by *one-way unification* (*pattern matching*). In conjunction with nesting, generic rules provide useful logical and associative capabilities, including good support for emulating arithmetic with natural numbers (base one) and usual data structures (such as lists, strings, and associative arrays). Recall that *instantiations* assign values to variables—ground values in pattern matching—whereas *unifications* are matching instantiations.

Leveraging their capabilities, most P systems variants, including cP systems, are able to transform "brute-force" algorithms into theoretically efficient solutions, with typically linear or sublinear runtimes. This allows the design of theoretically fast solutions to hard problems. Moreover, cP systems solutions for hard problems are typically the fastest, having small runtime coefficients. Additionally, cP systems solutions typically use small rulesets of fixed sizes, which do not change with the problem size (no uniform families, no polynomial preprocessing).

In this section, we introduce the basic features of a simplified version of cP systems, called *single-cell cP systems*, which have one single top-level cell, with nested sub-cells (thus there is no place for top-level cell networks and messaging). Listing 1 describes the basic formal syntax of single-cell cP-systems; for a more comprehensive description and explanation of cP systems, the reader is referred to [14,15]. This formal description consists of two BNF-like grammars, presented together, because of their similarities: (1) a *top-level cell*, in the sequel called *top-cell* (for brevity); (2) a *multiset rewriting rule*. Note that, in this figure and the sequel, we use the following two common abbreviations: *lhs* = left-hand-side, *rhs* = right-hand-side.

Listing 1. Simplified syntax for single-cell cP systems. Lhs = left-hand-side, rhs = right-hand-side, var-X = X may contain variables. Braces ({,}) and brackets ([,]) are meta-syntactic constructs followed by repetition bounds; here, braces generate *multisets*, whereas brackets generate *sequences*.

<top-cell> ::= <state> <objects>
<state> ::= <atom>
<objects> ::= {<atom> | <sub-cell>}$_0^\infty$
<sub-cell> ::= <functor> '('<objects> [';' <objects>]$_0^\infty$ ')'
<functor> ::= <atom>

. . . .

<rule> ::= <lhs> $\rightarrow_{\langle mode \rangle}$ <rhs> ['|' <promoters>]$_0^\infty$
<mode> ::= '1' | '+'
<lhs> ::= <state> <var-objects>
<rhs> ::= <state> <var-objects>
<state> ::= <atom>
<promoters> ::= <var-objects>
<var-objects> ::= {<variable> | <atom> | <var-sub-cell>}$_0^\infty$
<var-sub-cell> ::= <functor> '('<var-objects> [';' <var-objects>]$_0^\infty$ ')'
<functor> ::= <atom>

A single-cell cP system consists of one single *top-cell*, which—following the first grammar presented in Listing 1—has a *state* and contains *objects*, i.e., *atoms* and recursively nested *sub-cells*.

Remark 1. *In Prolog terminology, cell objects are terms, sub-cells are compound terms; and all cell objects are ground, i.e., cannot contain variables. Furthermore, unlike Prolog, cP functors do not have arities, and just represent multiset labels.*

Conventionally, atoms are represented by lowercase letters and variables by uppercase letters. A dedicated atom *1* is typically used to represent unary natural numbers (more details below). Anonymous (discard) variables in cP systems are denoted by underscores (_). The empty multiset is denoted by λ. As usual, multiset elements can be written in any order, and repetitions can be denoted as powers. Sample ground sub-cells: $a(bbc) = a(b^2c) = a(bcb), a(b(cc)\ d(ef)), n(111) = n(1^3)$.

Remark 2. *The grammar given in Listing 1 specifies that a sub-cell functor can be followed by a sequence of multiset arguments, which seems to require an ad hoc ordering concept. Functors with one single multiset argument are indeed essential in cP systems (similar to terms in Prolog), but functors with two or more arguments are not, because these could be replaced by one more level deeper cell nesting. For example, the sub-cell* $a(bc; de; fg)$ *could be also considered a shorthand for*

$a(bc \cdot (de) : (fg))$ (or $a(: (fg)\ bc \cdot (de))$, etc.), where the nested functors $(\cdot)$ and $(:)$ could be given ad hoc or provided by the system. Briefly, this conceptually redundant ordering appears for convenience *only, and the given grammar could be simplified, and strictly restricted to* nested labelled multisets. *Note that alternative definitions of cP systems use additional parentheses instead of the semicolons used here, e.g., the following two notations describe the same abstract syntax $a(bc;de;fg) \equiv a(bc)(de)(fg)$.*

As mentioned, *natural numbers* can be emulated using a dedicated unary symbol, such as *1*. By convention, we can also directly use the corresponding numbers, rather than their lower-level unary representation. For example:

$$111 = 1^3 = 3$$
$$\lambda = 1^0 = 0$$

A single-cell cP system evolves through a sequence of *configurations* by changing its *state* and *contents*. These changes are driven by the high-level *rewriting rules* associated to its top-cell, which are constructed according to the second grammar presented in Listing 1. Unlike similar cells in cell-like P systems, cP sub-cells are more restricted, by not having their own rules. Thus, sub-cells are just data storage facilities, and are acted upon by the top-cell's rules only. This restriction seems substantially outweighed by the extra power of the cP rules. Unlike other P systems variants, rules in cP systems are *generic* templates, i.e., their var-objects may contain *variables* that must be *instantiated* before the rule application.

Before a rule can apply:

- Its lhs state must match the current top-cell state.
- Its rhs state must match the already committed next state, if any, as further detailed below, in the section on weak priority order.
- The rule must be completely instantiated, i.e., all its variables must be replaced by ground objects, ensuring that its lhs and promoter var-objects match extant top-cell objects.

Rules are applied in a *weak priority order*, with rules considered in the given top-down order. Conventionally, the first lhs state is the state of the initial configuration. Once an applicable rule has been found, this commits to the next state, with subsequent rules committing to different states disabled. Rules going to the same state as the applicable rule, which can also be applied, will be applied in the *same step*. This state-based weak priority order supports a straightforward emulation of basic control flow (e.g., goto, conditional goto, or loop structures). Note that rules can be partitioned by their lhs state, without altering the semantics, as long as we keep the relative top-down order of rules starting with the same lhs state.

Essentially, applying a rule:

- Commits to the next state.
- Consumes (deletes) extant top-cell objects matching its lhs. Promoters must also match extant top-cell objects, but are not consumed by the rule.
- Creates new objects as indicated by its instantiated rhs. Newly created objects are temporary unavailable and become available after the end of the current step only, as in traditional P systems.

There are two rule application modes: exactly-once ($\rightarrow_1$) and max-parallel ($\rightarrow_+$). An *exactly-once* rule will apply for one single matching (*non-deterministically* chosen). A *max-parallel* rule will apply it as many times as possible, conceptually all in the same step, but following a *serialisation* semantics, i.e., its effects must be identical to a sequential repetition of the same rule in the exactly-once mode (sequence *non-deterministically* chosen). Although, as just mentioned, the cP semantics allow non-deterministic computations, most of our work has focused on *confluent* evolutions, often *deterministic*; the solution proposed in this paper is deterministic.

As with most other P system variants, the *runtime* of single-cell cP systems is measured in *steps*. Generally, a step is indicated by a state change, when a rule commits to a rhs state that differs from its lhs state. If the last applied rule does not change the state, then the control resumes at the first rule of that state, and this is also counted as a step. The system *halts* if a rule commits to a state with no associated rule; such states are called *final*. The system also halts if there are rules for the current state, but none is applicable (†). This last case, marked by a dagger (†), can be easily avoided by adding an extra catch-all rule, which will ensure termination in final states only.

As mentioned, like many other P system rules, cP rules have a significant potential for *non-determinism*. However, well-designed practical applications are highly deterministic. A cP system is *rule-deterministic* if each rule ends with exactly the same results, regardless of whether it is exactly-once or max-parallel, or how exactly it is instantiated and executed. A cP system is *step-deterministic* if each step is *locally confluent* with a guaranteed join after all step rules are applied, i.e., the step ends with exactly the same result, regardless of how its rules are applied. Obviously, rule-determinism is the stronger version, implying the weaker version, step-determinism. In both cases, we consider only evolutions that start from an expected initial configuration (not from arbitrary contents).

2.3. Examples

We provide several examples to clarify how cP systems are defined and used.

Example 3. *Matching examples, var-object (left) = ground object (right):*

- *Matching* $a(b(X)\ c(1X)) = a(b(1^2)\ c(1^3))$ *deterministically instantiates one single unifier:* $X = 1^2$.
- *Matching* $a(b(X)\ c(1X)) = a(b(1^2)\ c(1^2))$ *fails.*
- *Matching* $a(XY^2) = a(de^2f)$ *deterministically instantiates one single set of unifiers:* $X, Y = df, e$.
- *Matching* $a(XY) = a(df)$ *non-deterministically instantiates one of the following four sets of unifiers:* $X, Y = \lambda, df$; $X, Y = df, \lambda$; $X, Y = d, f$; $X, Y = f, d$.

Example 4. *Consider a cell in state* s_1 *that contains two objects* $a(1), a(11)$. *Depending on the actual application mode* $\alpha \in \{1, +\}$, *the following rule increments one or both a's by 1:*

$$s_1\ a(X) \rightarrow_\alpha s_2\ a(1X)$$

By unifying the lhs $a(X)$ against the given *a*s, two ground rules are instantiated:

$$s_1\ a(1) \rightarrow_1 s_2\ a(11) \quad (1)$$
$$s_1\ a(11) \rightarrow_1 s_2\ a(111) \quad (2)$$

When the application mode of the rule is exactly-once, $\alpha = 1$, the system non-deterministically applies one of the above two instantiations, (1) or (2). Thus, the result can be either $a(11), a(11)$ or $a(1), a(111)$.

However, when the application mode is max-parallel, $\alpha = +$, both instantiations are applied, and the result will be $a(11), a(111)$. Here, this transformation is rule-deterministic, not depending on the application order, (1,2) or (2,1).

Example 5. *Consider a cell in state* s_1 *that contains two objects* $a(1^3), b(1^5)$, *which respectively represent the numbers 3 and 5. The following rule destructively computes their sum,* $c = a + b$:

$$s_1\ a(X)\ b(Y) \rightarrow_1 s_2\ c(XY)$$

This rule is instantiated as $s_1\ a(1^3)\ b(1^5) \rightarrow_1 s_2\ c(1^8)$. Its application consumes the given $a(1^3)$ and $b(1^5)$, and creates a new objects $c(1^8)$, corresponding to the sum $3 + 5$.

Alternatively, a non-destructive summing can be performed using the following rule, where the given *a* and *b* appear as promoters:

$$s_1 \lambda \rightarrow_1 s_2\, c(XY) \mid a(X)\, b(Y)$$

Example 6. *Consider a cell in state s_1 that contains two objects $a(1)$ and three objects $b(11)$. The following max-parallel rule (1) consumes two $a(1)$ and two $b(11)$, creating two objects $c(111)$ and leaving exactly one $b(11)$; while the following max-parallel rule (2), which uses promoters, creates six objects $c(111)$, leaving the given as and bs intact:*

$$s_1\, a(X)\, b(Y) \rightarrow_+ s_2\, c(XY) \quad (1)$$
$$s_1\, \lambda \rightarrow_+ s_2\, c(XY) \mid a(X)\, b(Y) \quad (2)$$

Rule (1) could be considered rule-deterministic only if special configurations are guaranteed, such as the one given above; but, more generally, it is highly non-deterministic. Rule (2) is always rule-deterministic; essentially, it makes a Cartesian product of the given *as* and *bs*, concatenating the contents of all pairs.

Example 7. *Consider a cell in state s_1 that contains two objects $a(1)$ and three objects $a(11)$. The following max-parallel rule removes all duplicates, leaving exactly one $a(1)$ and one $a(11)$:*

$$s_1\, a(X) \rightarrow_+ s_2\, \lambda \mid a(X)$$

The application of this rule is equivalent to the following sequence of instantiations:

$$s_1\, a(1) \rightarrow_1 s_2\, \lambda \mid a(1)$$
$$s_1\, a(11) \rightarrow_1 s_2\, \lambda \mid a(11)$$
$$s_1\, a(11) \rightarrow_1 s_2\, \lambda \mid a(11)$$

The transformation is confluent, and the results will be the same, not depending on the relative application order of the above instantiations. After these three applications, no further unifying instantiations are possible because there are no longer sufficient remaining *as* to satisfy both the lhs and the promoter. Thus, this rule is rule-deterministic.

Example 8. *Consider a cell in state s_1 that contains one $a(...)$ and one $b(...)$, with unspecified contents. The following two-rule sequence models a non-destructive if-then-else operation, c = **if** $a \leq b$ **then** 0 **else** 1, accompanied by a state change (to either s_2 or s_3):*

$$s_1\, \lambda \rightarrow_1 s_2\, c(0) \mid a(X)\, b(X\_) \quad (1)$$
$$s_1\, \lambda \rightarrow_1 s_3\, c(1) \quad (2)$$

Rules are applied in weak-priority order. If rule (1) applies, then it commits to the target state s_2, so rule (2) becomes inapplicable. Otherwise, if rule (1) does not apply, the target state is still undecided, so rule (2) unconditionally applies and commits to target state s_3.

Example 9. *Consider a cell in state s_1 that contains a multiset of as, with numerical contents, e.g., $a(5), a(3), a(5), a(9), a(7)$. The following two max-parallel rules find the minimum in exactly two steps, regardless of the cardinality of the given multiset:*

$$s_1\, \lambda \rightarrow_+ s_2\, b(X) \mid a(X) \quad (1)$$
$$s_2\, b(X\_1) \rightarrow_+ s_3\, \lambda \mid a(X) \quad (2)$$

Rule (1) makes temporary working copies of all *as* as *bs*. Rule (2) deletes all *bs* for which there is a strictly lesser *a*. At state s_3, the cell contains one or more *bs*, all containing the same minimum value; in our given sample scenario, there will be one single $b(3)$. Both rules (1) and (2) are rule-deterministic.

3. The Logarithmic cP SAT Solution

We gradually develop our single-cell cP solution solution in three main phases. First, we show how a cP system can efficiently build all branches of a complete binary tree—this forms the backbone of our SAT solution. Secondly, we refine the building rules to decorate all these branches with explicit variable allocations; although conceptually redundant, explicit variable allocations are critical for efficient processing. Thirdly, and finally, we use these decorated tree branches to evaluate the given CNF formula, for all sets of variable allocations, which solves the SAT problem.

Leveraging the cP max-parallel mode, the full solution ruleset runs very efficiently, in $\mathcal{O}(\log n)$ time. It also has a small fixed size (25 rules) that does not depend on the problem size n (no uniform family, no polynomial preprocessing).

3.1. Building Trees

In this section we solve a subproblem that will later be incorporated in our SAT solution. Using a *single-cell cP system*, we aim to build a complete binary tree of height n, in deterministic $\mathcal{O}(\log n)$ time, by building its 2^n tree branches as cP objects. For simplicity, we also assume that n is a power of 2, $n = 2^k$, for some $k \geq 1$. If the given n is not a power of 2, we take n to be the next power of 2; we may thus obtain a bigger tree, which, however, does not affect our sought results.

The rules are shown in Listing 2. This ruleset has 8 rules, using 5 states, and assumes that n is given at the start via a namesake functor (e.g., $n(4)$). If needed, the reader is advised to crosscheck the appendix for an equivalent pseudocode, cf., Appendix B.

Listing 2. Ruleset for building complete binary trees of size n.

$$s_1\ \lambda \rightarrow_1 s_2\ h(1)\ t(\lambda;0)\ t(\lambda;1) \tag{1}$$

$$s_2\ h(N\_) \rightarrow_1 s_5\ \lambda \mid n(N) \tag{2}$$

$$s_2\ \lambda \rightarrow_+ s_3\ t'(X;Y) \mid t(X;Y) \tag{3}$$

$$s_3\ \lambda \rightarrow_+ s_4\ t''(t(X;Y);\ t(X';Y')) \mid t(X;Y)\ t'(X';Y') \tag{4}$$
$$s_3\ t(\_;\_) \rightarrow_+ s_4\ \lambda \tag{5}$$
$$s_3\ t'(\_;\_) \rightarrow_+ s_4\ \lambda \tag{6}$$

$$s_4\ t''(X;Y) \rightarrow_+ s_2\ t(X;Y) \tag{7}$$
$$s_4\ h(H) \rightarrow_1 s_2\ h(HH) \tag{8}$$

Rule (1) creates the starting tree, of height 1, with two branches. The current tree height is given by a sub-cell with functor h. Our tree branches are sub-cells with functor t and two arguments (two for consistency with the next branches that will be built via conceptual concatenation). The initial two branches are encoded as $t(\lambda;0)$ and $t(\lambda;1)$; by discarding the functors and parentheses, these encodings map to usual bit string labels, here 0 and 1, respectively. The cP encoding may seem to be overkill, but is required as cP systems lack strings, and are essentially based on amorphous multisets, where nesting is the only facility for structuring objects. For simplicity, in discussions, we will also use t as the name of the current tree (as the tree is completely defined by its branches).

Next, we repeatedly extend the current tree t, $k = \log n$ times (taking the ceiling if n is not power of 2), by transforming each leaf into the root of a new subtree t', ad hoc created as a structurally identical copy of t. Thus, the height of our trees grows exponentially: $1, 2, 4, 8, \ldots, 2^k = n$.

Rules (2–8) form the core loop of our system, starting at state s_2 and exiting at state s_5. Rule (2) breaks the loop if the current height h has reached (or exceeded) the given n.

Otherwise, rule (3) copies the current tree t into a temporary template t'. We note that the copy t' is not really needed here, but adds clarity.

Rules (4–6) creates the new higher tree t'', as the Cartesian product between the branches of t with the branches of t', then cleans the no-longer-needed objects t and t'. Each new branch is a concatenation of two previous branches and is represented as a new object t with two arguments, one for each component branch. For example, concatenating the branch z with the branch z' creates a new branch $t(z; z')$.

Rules (7–8) rename t'' as t, double the height h, and restart the loop from state s_2.

The following table lists the successively created branches, for $n = 4$. Anecdotally, note the relations $h = 2^k, d = k + 2$, where h is the height of the current tree t; k counts the completed iterations of loop (2–8); and d is the nesting depth of the branches.

k	h	branches–as bit strings	branches–as cP encodings
0	1	0; 1	$t(\lambda, 0);\ t(\lambda, 1)$
1	2	00; 01; 10; 11	$t(t(\lambda, 0), t(\lambda, 0));\ \ldots;\ t(t(\lambda, 1), t(\lambda, 1))$
2	4	0000; 0001; ...; 1111	$t(t(t(\lambda, 0), t(\lambda, 0)), t(t(\lambda, 0), t(\lambda, 0)));\ \ldots$

Theorem 1. *The cP ruleset 2 builds all branches of the complete binary tree of height n, in* $\mathcal{O}(\log n)$ *time.*

Proof. The previous discussion of the rules shows that they indeed build a complete binary tree. Rule (1) takes one step ($s_1 \to s_2$) and creates the initial complete binary tree of height $h = 1$. The loop formed by rules (2–8) takes 3 steps ($s_2 \to s_3 \to s_4 \to s_2$), runs $k = \log n$ times ($\lceil \log n \rceil$ times, if n is not power of 2), each time doubling the tree height h. The Cartesian product ensures that all created trees are still complete. The final break exit at rule (2) takes one more step ($s_2 \to s_5$). The total step count is $1 + 3 \log n + 1 = \mathcal{O}(\log n)$. The final height is $2^k = n$. □

Remark 3. *Ruleset 2 is rule-deterministic (and therefore also step-deterministic). Regardless of how it is instantiated and performed, each rule, whether exactly-once or max-parallel, ends with exactly the same results.*

3.2. Decorating Trees with Variable Allocations

In this section we extend the ruleset from the previous Section 3.1, by decorating all branches t with attributes a, representing explicit variable allocations. Although explicit allocations are, at first glance, redundant, because allocations can be recovered by parsing the branch label, they are critical for fast processing.

For example, looking at Figure 1, branch x should be decorated by allocations set $a(x)$, as follows: (i) for the height 1 tree: $a(0) = \{x_1 = 0\}$, $a(1) = \{x_1 = 1\}$; (ii) for the height 2 tree: $a(00) = \{x_1 = 0, x_2 = 0\}$, $a(01) = \{x_1 = 0, x_2 = 1\}$, etc. Furthermore, for a tree of height $4 = 2 + 2$, we should have $a(0100) = \{x_1 = 0, x_2 = 1, x_3 = 0, x_4 = 0\}$. Note that $a(0100) = a(01) \cup a'(00)$, where $a'(00) = \{x_{1+2} = 0, x_{2+2} = 0\}$, i.e., $a'(00)$ is $a(00)$ transformed by shifting the indices of its variables by +2.

Recalling that we build trees by means of successive concatenations, our ruleset formalises this intuition. Formally, the allocation set for branch $t(X; Y)$ is given by all sub-cells $a(X; Y; I; V)$, where I is a variable index and V its value (0 or 1). These a subsets are only virtually grouped together, solely by their shared branch label. This will not be a problem in regard to the logical and associative powers of cP systems. On the contrary, as we will see in the next Section 3.3, these loose associative collections will enable very fast evaluations.

The rules are shown in Listing 3. This ruleset has 14 rules, uses 6 states, and assumes that n is given at the start via a namesake functor (e.g., $n(4)$). If needed, the reader is advised to crosscheck the appendix: the sample traces listed in Appendix A and an equivalent pseudocode in Appendix C.

Listing 3. Ruleset for decorating trees.

$$s_1\ \lambda \rightarrow_1 s_2\ h(1)\ t(\lambda;0)\ t(\lambda;1)\ a(\lambda;0;1;0)\ a(\lambda;1;1;1) \tag{1}$$

$$s_2\ h(N\_) \rightarrow_1 s_6\ \lambda \mid n(N) \tag{2}$$

$$s_2\ \lambda \rightarrow_+ s_3\ t'(X;Y) \mid t(X;Y) \tag{3}$$
$$s_2\ \lambda \rightarrow_+ s_3\ a'(X;Y;IH;V) \mid h(H)\ a(X;Y;I;V) \tag{4}$$

$$s_3\ \lambda \rightarrow_+ s_4\ t''(t(X;Y);t(X';Y')) \mid t(X;Y)\ t'(X';Y') \tag{5}$$
$$s_3\ t(\_;\_) \rightarrow_+ s_4\ \lambda \tag{6}$$
$$s_3\ t'(\_;\_) \rightarrow_+ s_4\ \lambda \tag{7}$$

$$s_4\ \lambda \rightarrow_+ s_5\ a''(t(X;Y);Z;I;V) \mid t''(t(X;Y);Z)\ a(X;Y;I;V) \tag{8}$$
$$s_4\ \lambda \rightarrow_+ s_5\ a''(Z;t(X;Y);I';V) \mid t''(Z;t(X;Y))\ a'(X;Y;I';V) \tag{9}$$
$$s_4\ a(\_;\_;\_;\_) \rightarrow_+ s_5\ \lambda \tag{10}$$
$$s_4\ a'(\_;\_;\_;\_) \rightarrow_+ s_5\ \lambda \tag{11}$$

$$s_5\ t''(X;Y) \rightarrow_+ s_2\ t(X;Y) \tag{12}$$
$$s_5\ a''(X;Y;I;V) \rightarrow_+ s_2\ a(X;Y;I;V) \tag{13}$$
$$s_5\ h(H) \rightarrow_1 s_2\ h(HH) \tag{14}$$

Rule (1) creates the initial height 1 tree t and its allocations a (as mentioned above).

Rules (2–14) form the core loop, starting at state s_2 and exiting at state s_6. Rule (2) breaks the loop if $h \geq n$. Otherwise, rule (3) copies the current tree t into a temporary template t', and rule (4) copies the current allocations a into temporary objects a', *shifting* the variable indices by h.

Rules (5–7) creates the new higher tree t'', as the Cartesian product between the branches of t with the branches of t', then cleans the no-longer-needed objects t and t'. Rules (8,9) creates the allocations a'' for the new tree t'': rule (8) "lifts" the allocations a belonging to the former tree t, and rule (9) "lifts" the allocations a' belonging to the former template tree t'. Rules (10,11) clean the now-unneeded objects a and a'.

Rules (12–14) rename t'' as t and a'' as a, double the height h, and restart the loop from state s_2.

Arguments similar to those used in the proof of Theorem (1) lead us to the following result.

Proposition 1. *The cP ruleset 3 builds all branches of the complete binary tree of height n and decorates these with explicit variable allocations, in $\mathcal{O}(\log n)$ time.*

Remark 4. *Like its base, ruleset 2, ruleset 3 is rule-deterministic (and therefore also step-deterministic. Regardless of how it is instantiated and performed, each rule, whether exactly-once or max-parallel, ends with exactly the same results.*

3.3. Formula Evaluations

Up to this stage, the tree construction has ignored the actual problem, considering only its size and the number of variables, n. It is now time to introduce the formula that we actually want to solve. For this, we assume that the formula is given as the multiset of all its literal objects, r, where each literal object has the format $r(k;i;s)$, where k is a clause index in $[1,m]$, i is a variable index in $[1,n]$ and s is a sign in $\{-,+\}$, which indicates whether the clause k variable x_i is negated ($-$) or not ($+$).

For example, the formula of Example 1, $(x_1 \vee x_2) \wedge (\overline{x_1} \vee \overline{x_2})$, can be given as the multiset containing the following four r objects:

$$r(1;1;+)\ r(1;2;+)\ r(2;1;-)\ r(2;2;-)$$

For fast processing, we use a lookup table that quickly indicates the value of a literal, based on the variable value, regardless of whether or not the variable is negated. This lookup table is given by the following set with four w objects:

$w(0;+;0)\ w(0;-;1)\ w(1;+;1)\ w(1;-;0)$

where in $w(u;s;v)$, u is a variable value, s is a sign associated with a possible negation, and v is the literal value after considering s.

The rules are shown in Listing 4. This ruleset has 11 rules, uses 6 states, and assumes: (i) the r literal objects representing the given formula; (ii) the t and a objects as built by the ruleset of Listing 3. If needed, the reader is advised to crosscheck the appendix: the sample traces listed in Appendix A and an equivalent pseudocode in Appendix D.

Listing 4. Ruleset for formula evaluations (continuing from Ruleset 3).

$$s_6\ \lambda \rightarrow_+ s_7\ f(X;Y;K;I;S) \mid t(X;Y)\ r(K;I;S) \quad (15)$$

$$s_7\ f(X;Y;K;I;S) \rightarrow_+ s_8\ f'(X;Y;K;W) \mid a(X;Y;I;V)\ w(V;S;W) \quad (16)$$

$$s_8\ f'(X;Y;K;\_) \rightarrow_+ s_9\ \lambda \mid f'(X;Y;K;1) \quad (17)$$
$$s_8\ f'(X;Y;K;\_) \rightarrow_+ s_9\ \lambda \mid f'(X;Y;K;0) \quad (18)$$
$$s_8\ f'(X;Y;K;W) \rightarrow_+ s_9\ f''(X;Y;K;W) \quad (19)$$

$$s_9\ f''(X;Y;\_;\_) \rightarrow_+ s_{10}\ \lambda \mid f''(X;Y;\_;0) \quad (20)$$
$$s_9\ f''(X;Y;\_;\_) \rightarrow_+ s_{10}\ \lambda \mid f''(X;Y;\_;1) \quad (21)$$
$$s_9\ f''(X;Y;\_;W) \rightarrow_+ s_{10}\ f'''(X;Y;W) \quad (22)$$

$$s_{10}\ f'''(X;Y;\_) \rightarrow_+ s_{11}\ \lambda \mid f'''(X;Y;1) \quad (23)$$
$$s_{10}\ f'''(X;Y;\_) \rightarrow_+ s_{11}\ \lambda \mid f'''(X;Y;0) \quad (24)$$
$$s_{10}\ f'''(\_;\_;W) \rightarrow_1 s_{11}\ d(W) \quad (25)$$

The evaluation ruleset starts from s_6, the end state of the ruleset of Listing 3. Rule (15) makes a Cartesian product of branches and literals, for each branch t and literal r, creating an object f, which combines the branch t and the literal r.

Rule (16) transforms objects f into objects f', by replacing sign positions with actual literal values, taken from lookup table w. Briefly, these transformed f' objects record *evaluated literals*, separately for each branch and clause.

For each branch and clause, if there is a literal value 1, then rule (17) keeps this f' and deletes all other f' objects. Otherwise, if there still exists a literal value 0 (i.e., if all literal values were 0), then rule (18) keeps this f' and deletes all other f' objects (for the same branch and clause). At this stage, for each branch and clause, there is one single f' object left, indicating the clause value, 1 or 0. Rule (19) transforms these surviving f' objects into f'' objects, discarding the now-superfluous variable index. In a nutshell, f'' objects record *evaluated clauses*, separately for each branch.

Essentially, rules (20–22) repeat the same pattern and create f''' objects, which indicate *formula values*, separately for each branch. Now, if there is a branch where the formula is evaluated to 1, then rule (23) keeps this f''' and deletes all other f''' objects; otherwise, rule (24) keeps one single f''' that indicates 0 and deletes all other f''' objects.

Finally, there is exactly one f''' object left, which indicates whether or not there is an allocation that satisfies the formula. Using this sole surviving f''', rule (25) creates a d object that records the final decision.

Example 10. *The following table summarises the essential evaluation steps, in symbolic representation, for the formula of Example 1, cf. also Figure 1. Each branch has its own copy of formula literals, clause 1:* $\{x_1, x_2\}$*, clause 2:* $\{\overline{x_1}, \overline{x_2}\}$.

Branch	*Allocations*	*Eval. literals*	*Eval. clauses*	*Eval. formula*
00	$x_1 = 0, x_2 = 0$	$\{0,0\},\{1,1\}$	*0, 1*	*0*
01	$x_1 = 0, x_2 = 1$	$\{0,1\},\{1,0\}$	*1, 1*	*1*
10	$x_1 = 1, x_2 = 0$	$\{1,0\},\{0,1\}$	*1, 1*	*1*
11	$x_1 = 1, x_2 = 1$	$\{1,1\},\{0,0\}$	*1, 0*	*0*

If required, we could also return the set of all successful allocations, if any, but here we merely return the sought decision result, $d(0)$ (i.e., no), or $d(1)$ (i.e., yes). In our example case, there are two successful allocations, for branches 01 and 10, so the final decision is yes, $d(1)$.

Straightforward arguments show that the formula is exhaustively evaluated, and the evaluation ruleset takes a constant number of steps (5).

Proposition 2. *Given the complete binary tree built and decorated via the ruleset of Listing 3, the ruleset of Listing 4 solves SAT in $\mathcal{O}(1)$ time.*

Remark 5. *Ruleset 4 is only step-deterministic, not rule-deterministic. Three of its steps have deterministic step results, but consist of locally confluent fragments: 17–19, 20–22, and 23–25. The ruleset could be slightly modified to be strictly rule-deterministic, but we prefer the current version, due to its better readability.*

Noting that $\mathcal{O}(\log n) + \mathcal{O}(1) = \mathcal{O}(\log n)$, the following theorem is a direct consequence of Propositions 1 and 2. We also include a couple of static metrics provided by a close inspection of the rulesets of our two parts.

Theorem 2. *The SAT decision problem can be solved in $\mathcal{O}(\log n)$ time by means of a cP system ruleset with 11 states and 25 rules.*

3.4. Other NP-Complete Problems

Using the results of Stamm-Wilbrandt [23], Henderson et al. [4] have designed a cP solution that achieves a *constant time reduction*, $\mathcal{O}(1)$, from another famous NP-complete problem, k-colouring, to SAT. Combined with their *square root* SAT solution, $\mathcal{O}(\sqrt{n})$, they conclude that k-colouring and quite a few other NP-complete problems can be solved in *square root* time by cP-systems, as $\mathcal{O}(\sqrt{n}) + \mathcal{O}(1) = \mathcal{O}(\sqrt{n})$.

Based on the results of this paper, we similarly conclude that k-colouring, and possibly many other NP-complete problems, can be solved in *logarithmic* time by means of cP-systems, as $\mathcal{O}(\log n) + \mathcal{O}(1) = \mathcal{O}(\log n)$.

Theorem 3. *The k-colouring decision problem can be solved in $\mathcal{O}(\log n)$ time in the cP system model.*

4. Discussion

This section starts with a rough summary comparison of a few selected, and hopefully the most relevant, *deterministic* P systems solutions for the SAT problem. Essentially, we want to compare the *ruleset sizes* and the *running times*. Many of these solutions are linear, but their runtime often includes both the number of variables, n, and the number of clauses, m, e.g., $\mathcal{O}(m + n)$. See Nagy [6] for a short survey on some of the previous P system solutions.

There is also a recently proposed cP solution by Henderson et al. [4], which managed a remarkable breakthrough, being sublinear, $\mathcal{O}(\sqrt{n})$. Our new solution, proposed in this paper, shows that cP systems are able to solve SAT and other NP-complete problems in a substantially faster sublinear time of $\mathcal{O}(\log n)$. As seen in Table 1, our novel solution surpasses all other extant solutions in runtime, and is comparable to the number of template rules (more about this below).

This comparison is a difficult problem by itself, as the many P systems variants have substantial differences, so one should be careful when "comparing apples with oranges", and then drawing strong conclusions. First, all P systems measure the runtimes in terms of *steps*, which at the first seems to be a uniform measure, but the definition of steps may differ among variants, and may have different granularity.

Secondly, the rules also have different granularity. Here, we attempt to create a more level playing-field by following the methods used by Henderson et al. [16]. Thus, we indicate the ruleset size in two ways: (i) the actual number of *rules*, and (ii) the number of *rule templates*. As defined in [16], rule templates are groupings of similar rules, differing only by symbol indices, e.g., $a_i \to b_i, i = 1, 2, \ldots, n$, which represents n rules but one single rule template. This should considerably level the playing field, as such a template is typically subsumed by one single generic rule in cP systems, e.g., $a(I) \to b(I) \mid c(I)$.

On the other side, when counting rule templates and rules, we did not consider the numbers of repeated copies placed in different membranes/neurons. Additionally, non cP systems solutions are not single solutions, but uniform families of solutions, i.e., a different solution will be used for each different problem size, typically following the same templates, but with different alphabet and ruleset sizes. The needed pre-processing time was roughly estimated from the papers, and presented in a separate column. cP systems do not have such facilities, as they use a fixed ruleset that must be defined in the top-level cell only (subcells do not have their own rules). This may seem to create some bias against cP systems, but we feel that the power of generic rules will finally rebalance the comparison.

Table 1. Ruleset size and runtime for several proposed P system solutions. † = this paper. The preprocessing time was only estimated by us.

Paper	P System Variant	#Templates	#Rules	Runtime	Preprocessing
[7] 2006	with active membranes	27	$\mathcal{O}(mn^2)$	$\mathcal{O}(m+n)$	$\Theta(mn^2)$
[8] 2016	with proteins on membranes	22	$\mathcal{O}(mn)$	$\mathcal{O}(mn)$	$\Theta(mn)$
[9] 2017	tissue-like	29	$\mathcal{O}(mn^2)$	$\mathcal{O}(m+n)$	$\Theta(mn^2)$
[4] 2021	cP system	19	19	$\mathcal{O}(\sqrt{n})$	NA
† 2021	cP system	25	25	$\mathcal{O}(\log n)$	NA

We conclude this section by noting several research directions that could follow the current result. (1) Design a shallow solution for this problem. (2) As a combined method of space and task optimisation, partially evaluate the given formula while building the tree. This would enable one to prune branches that cannot lead to any solution, because one of the clauses is already false. This should substantially reduce the actual work, and balance it better, possibly leading to more efficient practical implementations. (3) Develop a similar approach for QSAT, a famous related PSPACE-complete problem, which is substantially more complex and challenging. (4) Investigate the feasibility of similar solutions in other P system variants.

5. Conclusions

In this work, we have presented a novel cP solution to SAT, a famous NP-complete (and thus NP-hard) problem. Our solution is deterministic and runs in logarithmic time, $\mathcal{O}(\log n)$. To the best of our knowledge, this represents a significant breakthrough in membrane computing, being orders-of-magnitude faster than all previous *deterministic* solutions.

In conjunction with a couple of known reduction results, our solution enables further logarithmic-time solutions, $\mathcal{O}(\log n)$, to other NP-complete problems, such as k-colouring.

Our results open the way to several other challenging research problems, such as extending this method to cover QSAT (which is a substantially harder, PSPACE-complete problem); designing a time- and space-optimised version and possibly a shallow version; and investigating the feasibility of similar solutions in other P system variants.

Author Contributions: Conceptualisation, investigation, and formal analysis: R.N. and A.H.; supervision: R.N. and M.J.D.; writing—original draft preparation: R.N., A.H., J.C., and Y.L.; writing—review and editing: R.N., M.J.D., A.H., J.C., and Y.L.; validation: Y.L. All authors have read and agreed to the published version of the manuscript.

Funding: This research received no external funding.

Institutional Review Board Statement: Not applicable.

Informed Consent Statement: Not applicable.

Data Availability Statement: Not applicable.

Conflicts of Interest: The authors declare no conflict of interest.

Appendix A. Traces for Sections "Decorating Trees with Variable Allocations" and "Ruleset Evaluations"

This section traces critical configuration fragments for the whole proposed SAT algorithm, i.e., the combined rulesets 3 and 4. The trace is organised by steps, listing essential configuration contents at the start of each new step. The initial configuration does not change, so it only appears for state s_1. We again consider the formula of Example 1: $(x_1 \vee x_2) \wedge (\overline{x_1} \vee \overline{x_2})$, with $n = 2, m = 2$.

For readability, the two components of nested cP branch labels, which appear as arguments for functors t, a, f (possibly primed), are indicated by their corresponding binary equivalents (cf,. Section 2.1), which are underlined, e.g.,: $t(\underline{0}) = t(\lambda;0)$, $t(\underline{1}) = t(\lambda;1)$, $t(\underline{00}) = t(t(\lambda;0);t(\lambda;0))$, $t(\underline{01}) = t(t(\lambda;0);t(\lambda;1))$, $a(\underline{01};1;0) = a(t(\lambda;0);t(\lambda;1);1;0)$, $f(\underline{01};1;2;+) = f(t(\lambda;0);t(\lambda;1);1;2;+)$, $f''(\underline{01};2;1) = f''(t(\lambda;0);t(\lambda;1);2;1)$, etc.

- Enter state s_1, with immutable objects (not further listed unless actually useful):

$n(2)$
$w(0;+;0)\ w(0;-;1)\ w(1;+;1)\ w(1;-;0)$
$r(1;1;+)\ r(1;2;+)\ r(2;1;-)\ r(2;2;-)$

- Step $s_1 \to s_2$, rule (1): Create initial height 1 tree objects, t and a.
- Enter state s_2, with:

$n(2)\ h(1)$
$t(\underline{0})\ t(\underline{1})\ a(\underline{0};1;0)\ a(\underline{1};1;1)$

- Step $s_2 \to s_3$, rules (3-4): Enter the loop, duplicate tree objects t and a, as t' and a'.
- Enter state s_3, with:

$h(1)$
$t(\underline{0})\ t(\underline{1})\ a(\underline{0};1;0)\ a(\underline{1};1;1)$
$t'(\underline{0})\ t'(\underline{1})\ a'(\underline{0};2;0)\ a'(\underline{1};2;1)$

- Step $s_3 \to s_4$, rules (5-7): Create double height tree t'' by the Cartesian product of t and t'.
- Enter state s_4, with:

$h(1)$
$t''(\underline{00})\ t''(\underline{01})\ t''(\underline{10})\ t''(\underline{11})$
$a(\underline{0};1;0)\ a(\underline{1};1;1)\ a'(\underline{0};2;0)\ a'(\underline{1};2;1)$

- Step $s_4 \rightarrow s_5$, rules (8-11): Create a'', allocation attributes for t''.
- Enter state s_5, with:

$$\begin{array}{l}
h(1)\\
t''(\underline{00})\ t''(\underline{01})\ t''(\underline{10})\ t''(\underline{11})\\
a''(\underline{00};1;0)\ a''(\underline{01};1;0)\ a''(\underline{10};1;1)\ a''(\underline{11};1;1)\\
a''(\underline{00};2;0)\ a''(\underline{10};2;0)\ a''(\underline{01};2;1)\ a''(\underline{11};2;1)
\end{array}$$

- Step $s_5 \rightarrow s_2$, rules (12–14): Double the height and rename temporary tree objects t'' and a'' as t and a.
- Enter state s_2, with:

$$\begin{array}{l}
n(2)\ h(2)\\
t(\underline{00})\ t(\underline{01})\ t(\underline{10})\ t(\underline{11})\\
a(\underline{00};1;0)\ a(\underline{00};2;0)\ a(\underline{01};1;0)\ a(\underline{01};2;1)\\
a(\underline{10};1;1)\ a(\underline{10};2;0)\ a(\underline{11};1;1)\ a(\underline{11};2;1)
\end{array}$$

- Step $s_2 \rightarrow s_6$, rule (2): Take loop exit.
- Enter state s_6 (end of ruleset 3, and start of 4), with:

$$\begin{array}{l}
r(1;1;+)\ r(1;2;+)\ r(2;1;-)\ r(2;2;-)\\
t(\underline{00})\ a(\underline{00};1;0)\ a(\underline{00};2;0)\\
t(\underline{01})\ a(\underline{01};1;0)\ a(\underline{01};2;0)\\
t(\underline{10})\ a(\underline{10};1;0)\ a(\underline{10};2;0)\\
t(\underline{11})\ a(\underline{11};1;0)\ a(\underline{11};2;0)
\end{array}$$

- Step $s_6 \rightarrow s_7$, rule (15): Multiply formula literals, making copies for each branch.
- Enter state s_7, with:

$$\begin{array}{l}
w(0;+;0)\ w(0;-;1)\ w(1;+;1)\ w(1;-;0)\\
t(\underline{00})\ a(\underline{00};1;0)\ a(\underline{00};2;0)\ f(\underline{00};1;1;+)\ f(\underline{00};1;2;+)\ f(\underline{00};2;1;-)\ f(\underline{00};2;2;-)\\
t(\underline{01})\ a(\underline{01};1;0)\ a(\underline{01};2;0)\ f(\underline{01};1;1;+)\ f(\underline{01};1;2;+)\ f(\underline{01};2;1;-)\ f(\underline{01};2;2;-)\\
t(\underline{10})\ a(\underline{10};1;0)\ a(\underline{10};2;0)\ f(\underline{10};1;1;+)\ f(\underline{10};1;2;+)\ f(\underline{10};2;1;-)\ f(\underline{10};2;2;-)\\
t(\underline{11})\ a(\underline{11};1;0)\ a(\underline{11};2;0)\ f(\underline{11};1;1;+)\ f(\underline{11};1;2;+)\ f(\underline{11};2;1;-)\ f(\underline{11};2;2;-)
\end{array}$$

- Step $s_7 \rightarrow s_8$, rule (16): Evaluate literals.
- Enter state s_8, with:

$$\begin{array}{l}
t(\underline{00})\ a(\underline{00};1;0)\ a(\underline{00};2;0)\ f'(\underline{00};1;0)\ f'(\underline{00};1;0)\ f'(\underline{00};2;1)\ f'(\underline{00};2;1)\\
t(\underline{01})\ a(\underline{01};1;0)\ a(\underline{01};2;0)\ f'(\underline{01};1;0)\ f'(\underline{01};1;1)\ f'(\underline{01};2;1)\ f'(\underline{01};2;0)\\
t(\underline{10})\ a(\underline{10};1;0)\ a(\underline{10};2;0)\ f'(\underline{10};1;1)\ f'(\underline{10};1;0)\ f'(\underline{10};2;0)\ f'(\underline{10};2;1)\\
t(\underline{11})\ a(\underline{11};1;0)\ a(\underline{11};2;0)\ f'(\underline{11};1;1)\ f'(\underline{11};1;1)\ f'(\underline{11};2;0)\ f'(\underline{11};2;0)
\end{array}$$

- Step $s_8 \rightarrow s_9$, rules (17–19): Disjunctions between literals.
- Enter state s_9, with:

$$\begin{array}{l}
t(\underline{00})\ a(\underline{00};1;0)\ a(\underline{00};2;0)\ f''(\underline{00};1;0)\ f''(\underline{00};2;1)\\
t(\underline{01})\ a(\underline{01};1;0)\ a(\underline{01};2;0)\ f''(\underline{01};1;1)\ f''(\underline{01};2;1)\\
t(\underline{10})\ a(\underline{10};1;0)\ a(\underline{10};2;0)\ f''(\underline{10};1;1)\ f''(\underline{10};2;1)\\
t(\underline{11})\ a(\underline{11};1;0)\ a(\underline{11};2;0)\ f''(\underline{11};1;1)\ f''(\underline{11};2;0)
\end{array}$$

- Step $s_9 \rightarrow s_{10}$, rules (20–22): Conjunctions between clauses.
- Enter state s_{10}, with:

$$\begin{array}{l}
t(\underline{00})\ a(\underline{00};1;0)\ a(\underline{00};2;0)\ f'''(\underline{00};0)\\
t(\underline{01})\ a(\underline{01};1;0)\ a(\underline{01};2;0)\ f'''(\underline{01};1)\\
t(\underline{10})\ a(\underline{10};1;0)\ a(\underline{10};2;0)\ f'''(\underline{10};1)\\
t(\underline{11})\ a(\underline{11};1;0)\ a(\underline{11};2;0)\ f'''(\underline{11};0)
\end{array}$$

- Step $s_{10} \rightarrow s_{11}$, rules (23–25): Disjunction between branches and final decision.
  - Intermediate snapshot after rules (23–24):

$$\begin{array}{|l|}\hline t(\underline{00})\ a(\underline{00};1;0)\ a(\underline{00};2;0) \\ t(\underline{01})\ a(\underline{01};1;0)\ a(\underline{01};2;0)\ f'''(\underline{01};1) \\ t(\underline{10})\ a(\underline{10};1;0)\ a(\underline{10};2;0) \\ t(\underline{11})\ a(\underline{11};1;0)\ a(\underline{11};2;0) \\ \hline\end{array}$$

- Enter state s_{11} (end, with success), with:

$$\begin{array}{|l|}\hline d(1) \\ t(\underline{00})\ a(\underline{00};1;0)\ a(\underline{00};2;0) \\ t(\underline{01})\ a(\underline{01};1;0)\ a(\underline{01};2;0) \\ t(\underline{10})\ a(\underline{10};1;0)\ a(\underline{10};2;0) \\ t(\underline{11})\ a(\underline{11};1;0)\ a(\underline{11};2;0) \\ \hline\end{array}$$

Appendix B. Pseudocode for Section "Building Trees"

The pseudocode is shown in Listing A1. We assume that n is already given as an initial parameter. Multisets are denoted by capital letters, e.g., T is the multiset (actually set) of all t objects. Branches are represented by their intuitive bit string notation (not as cP encodings). At state s_3, the Cartesian product ($\times$) is followed by projecting string concatenations ($\cdot$) of all pairs, which creates double-length branches.

Listing A1. Pseudocode for the ruleset of Listing 2.

s_1:
$h \leftarrow 1; T \leftarrow \{0,1\}$ — *// initial tree height and branches*

s_2:
if $h \geq n$ **then goto** s_5 **else** — *// alt* **while** $h < n$ **do** *...*

  $T' \leftarrow T$ — *// copy current branches*

s_3:
  $T'' \leftarrow \{t \cdot t' \mid (t,t') \in T \times T'\}$ — *// concatenate all branch pairs*
  $T \leftarrow$ **null**
  $T' \leftarrow$ **null**

s_4:
  $T \leftarrow T''; T'' \leftarrow$ **null** — *// next tree*
  $h \leftarrow h + h$ — *// next height*
  goto s_2

s_5: — *// end*

Appendix C. Pseudocode for Section "Decorating Trees with Variable Allocations"

The pseudocode is shown in Listing A2. We assume that n is already given as an initial parameter. Multisets are denoted by capital letters, e.g., T is the set of all t objects (branches), where branches are represented by their intuitive bit string notation (not as cP encodings).

Variable allocations are given as partial functions $[1,n] \rightarrow \{0,1\}$. For example, using a Python-like notation, the allocation set $\{x_1 = 0, x_2 = 1\}$ is represented as the list $\alpha = \{1 : 0, 2 : 1\}$; thus $\alpha[2] = 1$. At state s_2, σ is a transformation that *shifts* the variable indices in a given allocation set α by a given number h, i.e., $\sigma(\alpha,h) = \{i : (v+h) \mid (i : v) \in \alpha\}$; e.g., $\sigma(\{1:0,2:1\},2) = \{3:0,4:1\}$, and, more symbolically, $\{x_3 = 0, x_4 = 1\}$.

At state s_3, the Cartesian product ($\times$) is followed by projecting string concatenations ($\cdot$) of all pairs, which creates double-length branches.

Listing A2. Pseudocode for the ruleset of Listing 3.

```
s1:
h ← 1; T ← {0,1}                                   // initial tree height and branches
A ← {(0,{1 : 0}),(1,{1 : 1})}                      // initial branch variable allocations

s2:
if h ≥ n then goto s6 else                         // alt while h < n do ...

    T' ← T                                         // copy current branches
    A' ← {(t,σ(α,h)) | (t,α) ∈ A}                  // copy allocations and shift indices by h

s3:
    T'' ← {t · t' | (t,t') ∈ T × T'}               // concatenate all branch pairs
    T ← null
    T' ← null

s4:
    A'' ← {(t · t',α) | t · t' ∈ T'', |t| = |t'|, (t,α) ∈ A}     // lift from A
        ∪ {(t · t',α) | t · t' ∈ T'', |t| = |t'|, (t',α) ∈ A'}   // lift from A'
    A ← null
    A' ← null

s5:
    T ← T''; T'' ← null                            // next tree
    A ← A''; A'' ← null                            // next allocations
    h ← h + h                                      // next height
    goto s2

s6:                                                // end (of this phase)
```

Appendix D. Pseudocode for Section "Ruleset Evaluations"

The pseudocode is shown in Listing A3. We assume that this code follows the code of the preceding section, given in Listing A2. As before, $\times$ denotes the Cartesian product operator.

T is the set of all branches and A is the (conceptually redundant) set of all associated allocations. as constructed using the preceding pseudocode A2, R_k is the set of all literals that appear in clause $k \in [1, m]$, and R is the set of all possible literals. For example, assuming that its clauses are indexed in left-to-right order, the previously discussed formula, $(x_1 \vee x_2) \wedge (\overline{x_1} \vee \overline{x_2})$, is given by $R_1 = \{x_1, x_2\}$, $R_2 = \{\overline{x_1}, \overline{x_2}\}$, $R = \{x_1, x_2, \overline{x_1}, \overline{x_2}\}$.

We also assume a function $\omega : R \times A \rightarrow \{0,1\}$, roughly corresponding to our w lookup, such that $\omega(r, \alpha)$ is the Boolean value of literal r for the allocation set α (considering its possible negation). For example, assume that (in symbolical form): $r = x_1, r' = \overline{x_1}$; and $\alpha = \{x_1 = 0, x_2 = 1\}$, i.e., the symbolical form of $\{1 : 0, 2 : 1\}$. Then, $\omega(r, \alpha) = 0, \omega(r', \alpha) = 1$.

Listing A3. Pseudocode for the ruleset of Listing 4.

s_6: // *attach literal copies to each branch in T*
$F \leftarrow \bigcup_{k=1}^{m}(T \times \{k\} \times R_k)$ // *F is (normally) a set*

s_7: // *evaluate literals for branch t and clause k*
$F' \leftarrow \{(t,k,w) \mid (t,k,r) \in F, (t,\alpha) \in A, \omega(r,\alpha) = w\}$ // *take F′ as a multiset!*
$F \leftarrow$ **null**

s_8: // *evaluate each clause for branch t, using disjunctions between literals*
$F'' \leftarrow \{(t,k,1) \mid (t,k,1) \in F'\}$ // *take F″ as a set*
$\cup \{(t,k,0) \mid (t,k,1) \notin F'\}$
$F' \leftarrow$ **null**

s_9: // *evaluate formula for branch t, using conjunctions between clauses*
$F''' \leftarrow \{(t,0) \mid \exists k \in [1,m], (t,k,0) \in F''\}$ // *take F‴ as a set*
$\cup \{(t,1) \mid \forall k \in [1,m], (t,k,0) \notin F''\}$
$F'' \leftarrow$ **null**

s_{10}: // *the decision is yes, if there is at least one branch evaluating true*
$d \leftarrow$ **if** $\exists t \in T, (t,1) \in F'''$ **then** 1 **else** 0
$F''' \leftarrow$ **null**

s_{11}: // *end*

References

1. Sipser, M. *Introduction to the Theory of Computation*; Cengage Learning: Boston, MA, USA, 2012.
2. Baker, B.S. Approximation algorithms for NP-complete problems on planar graphs. *J. ACM (JACM)* **1994**, *41*, 153–180. [CrossRef]
3. Downey, R.G.; Fellows, M.R. Fixed-parameter tractability and completeness I: Basic results. *SIAM J. Comput.* **1995**, *24*, 873–921. [CrossRef]
4. Henderson, A.; Nicolescu, R.; Dinneen, M.J. *Sublinear P System Solutions to NP-Complete Problems*; CDMTCS Report 559; University of Auckland: Auckland, New Zealand, 2022. Available online: https://www.cs.auckland.ac.nz/research/groups/CDMTCS/researchreports/download.php?selected-id=831 (accessed on 14 January 2022).
5. Manca, V. DNA and membrane algorithms for SAT. *Fundam. Inform.* **2002**, *49*, 205–221.
6. Nagy, B. On efficient algorithms for SAT. In *International Conference on Membrane Computing*; LNCS 7762; Csuhaj-Varjú, E., Gheorghe, M., Rozenberg, G., Salomaa, A., Vaszil, G., Eds.; Springer: Berlin/Heidelberg, Germany, 2012; pp. 295–310. [CrossRef]
7. Pan, L.; Alhazov, A. Solving HPP and SAT by P systems with active membranes and separation rules. *Acta Inform.* **2006**, *43*, 131–145. [CrossRef]
8. Song, B.; Pérez-Jiménez, M.J.; Pan, L. An efficient time-free solution to SAT problem by P systems with proteins on membranes. *J. Comput. Syst. Sci.* **2016**, *82*, 1090–1099. [CrossRef]
9. Song, B.; Zhang, C.; Pan, L. Tissue-like P systems with evolutional symport/antiport rules. *Inf. Sci.* **2017**, *378*, 177–193. [CrossRef]
10. Păun, G. Computing with membranes. *J. Comput. Syst. Sci.* **2000**, *61*, 108–143. [CrossRef]
11. Păun, G. P systems with active membranes: Attacking NP-Complete problems. *J. Autom. Lang. Comb.* **2001**, *6*, 75–90. [CrossRef]
12. Martín-Vide, C.; Păun, G.; Pazos, J.; Rodríguez-Patón, A. Tissue P systems. *Theor. Comput. Sci.* **2003**, *296*, 295–326. [CrossRef]
13. Ionescu, M.; Păun, G.; Yokomori, T. Spiking neural P systems. *Fundam. Inform.* **2006**, *71*, 279–308.
14. Nicolescu, R.; Henderson, A. An Introduction to cP Systems. In *Enjoying Natural Computing: Essays Dedicated to Mario de Jesús Pérez-Jiménez on the Occasion of His 70th Birthday*; Graciani, C., Riscos-Núñez, A., Păun, G., Rozenberg, G., Salomaa, A., Eds.; LNCS 11270; Springer: Berlin/Heidelberg, Germany, 2018; pp. 204–227. [CrossRef]
15. Henderson, A.; Nicolescu, R. Actor-Like cP Systems. In *Membrane Computing*; Hinze T., Rozenberg G., Salomaa A., Zandron C., Eds.; LNCS 11399; Springer: Berlin/Heidelberg, Germany, 2019; pp. 160–187. [CrossRef]
16. Henderson, A.; Nicolescu, R.; Dinneen, M.J. Solving a PSPACE-complete problem with cP systems. *J. Membr. Comput.* **2020**, *2*, 311–322. [CrossRef]
17. Ishdorj, T.O.; Leporati, A.; Pan, L.; Zeng, X.; Zhang, X. Deterministic solutions to QSAT and Q3SAT by spiking neural P systems with pre-computed resources. *Theor. Comput. Sci.* **2010**, *411*, 2345–2358. [CrossRef]
18. Leporati, A.; Manzoni, L.; Mauri, G.; Porreca, A.E.; Zandron, C. Solving QSAT in sublinear depth. In *Membrane Computing*; Hinze, T., Rozenberg, G., Salomaa, A., Zandron, C., Eds.; LNCS 11399; Springer: Berlin/Heidelberg, Germany, 2019; pp. 188–201. [CrossRef]

19. Gutiérrez-Naranjo, M.A.; Pérez-Jiménez, M.J.; Romero-Campero, F.J. A Linear Solution for QSAT with Membrane Creation. In *Membrane Computing*; Freund, R., Păun, G., Rozenberg, G., Salomaa, A., Eds.; LNCS 3850; Springer: Berlin/Heidelberg, Germany, 2006; pp. 241–252. [CrossRef]
20. Alhazov, A.; Pérez-Jiménez, M.J. Uniform solution of QSAT using polarizationless active membranes. In *International Conference on Machines, Computations, and Universality*; Durand-Lose, J., Margenstern, M., Eds.; LNCS 4664; Springer: Berlin/Heidelberg, Germany, 2007; pp. 122–133. [CrossRef]
21. Leporati, A.; Manzoni, L.; Mauri, G.; Porreca, A.; Zandron, C. Characterizing PSPACE with shallow non-confluent P systems. *J. Membr. Comput.* **2019**, *1*, 75–84. [CrossRef]
22. Leporati, A.; Mauri, G.; Zandron, C.; Păun, G.; Pérez-Jiménez, M. Uniform solutions to SAT and Subset Sum by spiking neural P systems. *Nat. Comput.* **2009**, *8*, 681–702. [CrossRef]
23. Stamm-Wilbrandt, H. *Programming in Propositional Logic or Reductions: Back to the Roots (Satisfiability)*; Sekretariat für Forschungsberichte, Inst. für Informatik III, University of Bonn: Bonn, Germany, 1993.

MDPI
St. Alban-Anlage 66
4052 Basel
Switzerland
Tel. +41 61 683 77 34
Fax +41 61 302 89 18
www.mdpi.com

Axioms Editorial Office
E-mail: axioms@mdpi.com
www.mdpi.com/journal/axioms

www.ingramcontent.com/pod-product-compliance
Lightning Source LLC
LaVergne TN
LVHW070609100526
838202LV00012B/604